Applied Probability and Statistics (*Continued*)

continued on back

Statistical Methods for Rates and Proportions

A WILEY PUBLICATION IN APPLIED STATISTICS

Statistical Methods for Rates and Proportions

JOSEPH L. FLEISS

Professor and Head of the Division of Biostatistics at the Columbia University School of Public Health, 600 West 168th Street, New York, New York 10032

A WILEY-INTERSCIENCE PUBLICATION

JOHN WILEY & SONS   New York . London . Sydney . Toronto

Library of Congress Cataloging in Publication Data:

Fleiss, Joseph L
  Statistical methods for rates and proportions.

  (Wiley series in probability and mathematical statistics) (A Wiley Publication in mathematical statistics)
  Includes bibliographies.

  1. Analysis of variance. 2. Sampling (Statistics) 3. Biometry. I. Title. II. Title: Rates and proportions.

QA279.F58   519.5'35   72-8521
ISBN 0-471-26370-2

Printed in the United States of America

10 9 8 7 6 5

TO ISABEL

Preface

This book is concerned solely with comparisons of qualitative or categorical data. The case of quantitative data is treated in the many books devoted to the analysis of variance. Other books have restricted attention to categorical data (such as A. E. Maxwell, *Analysing qualitative data*, Methuen, London, 1961, and R. G. Francis, *The rhetoric of science: A methodological discussion of the two-by-two table*, University of Minnesota Press, Minneapolis, 1961), but an updated monograph seemed overdue. A recent text (D. R. Cox, *The analysis of binary data*, Methuen, London, 1970) is at once more general than the present book in that it treats categorical data arising from more complicated study designs and more restricted in that it does not treat such topics as errors of misclassification and standardization of rates.

Although the ideas and methods presented here should be useful to anyone concerned with the analysis of categorical data, the emphasis and examples are from the disciplines of clinical medicine, epidemiology, psychiatry and psychopathology, and public health. The book is aimed at research workers and students who have had at least a year's course in applied statistics, including a thorough grounding in chi square and correlation. Most chapters conclude with one or more problems. Some call for the proof of an algebraic identity. Others are numerical, designed either to have the reader apply what he has learned or to present ideas mentioned only in passing in the text.

No more complicated mathematical techniques than the taking of logarithms and the extraction of square roots are required to apply the methods described. This means that anyone with only high school algebra, and with only a desktop calculator, can apply the methods presented. It also means, however, that analyses requiring matrix inversion or other complicated mathematical techniques (e.g., the analysis of multiple contingency tables) are not described. Instead, the reader is referred to appropriate sources.

The estimation of the degree of association or difference assumes equal importance with the assessment of statistical significance. Except where the formulas are excessively complicated, I present the standard error of almost every measure of association or difference given in the text. The standard errors are used to test hypotheses about the corresponding parameters, to compare the precisions of different methods of estimation, and to obtain a weighted average of a number of independent estimates of the same parameter.

I have tried to be careful in giving both sides of various arguments that are still unresolved about the proper design of studies and analysis of data. Examples are the use of matched samples and the measurement of association. Inevitably, my own biases have probably affected how I present the opposing arguments.

In two instances, however, my bias is so strong that I do not even permit the other side to be heard. I do not find confidence intervals to be useful, and therefore do not discuss interval estimation at all. The reader who finds a need for confidence intervals will have to refer to some of the cited references for details. He will find, by the way, that the proper interval is almost always more complicated than simply the point estimate plus or minus a multiple of its standard error.

The second instance is my bias against the Bayesian approach to statistical inference. See W. Edwards, H. Lindman, and L. J. Savage, Bayesian statistical inference for psychological research, *Psychol. Rev.*, **70**, 193–242, 1963, for a description of the Bayesian approach to data in psychology; and J. Cornfield, A Bayesian test of some classical hypotheses—with applications to sequential clinical trials, *J. Amer. statist. Assoc.*, **61**, 577–594, 1966, for a description of that approach to data in medicine. I believe that the kind of thinking described in Chapter 3, especially in Section 3.1, provides an adequate alternative to the Bayesian approach.

It is with gratitude that I acknowledge the advice, criticism, and encouragement of Professors John Fertig, Mervyn Susser, and Andre Varma of Columbia University and of Dr. Joseph Zubin of the Biometrics Research unit of the New York State Department of Mental Hygiene. Dr. Gary Simon of Princeton University and Professor W. Edwards Deming of New York University reviewed the manuscript and pointed out a number of errors I had made in an earlier draft. Needless to say, I take full responsibility for any and all errors that remain.

My wife Isabel was a constant source of inspiration as well as an invaluable editorial assistant.

The major portion of the typing was admirably performed by Vilma Rivieccio. Additional typing, collating, and keypunching were ably carried out by Blanche Agdern, Rosalind Fruchtman, Cheryl Keller, Sarah Lichtenstaedter, and Edith Pons.

My work was supported in part by grant DR 00793 from the National Institute of Dental Research (John W. Fertig, Ph.D., Principal Investigator) and in part by grant MH 08534 from the National Institute of Mental Health (Robert L. Spitzer, M.D., Principal Investigator). Except when noted otherwise, the tables in the Appendix were generated by programs run on the computers of the New York State Psychiatric Institute and of Rockland State Hospital.

I thank Professor E. S. Pearson and the Biometrika Trustees for permission to quote from Tables 1, 4, and 8 of *Biometrika tables for statisticians, Vol. I,* edited by E. S. Pearson and H. O. Hartley; John Wiley & Sons for permission to use Tables A.1 to A.3 of *Statistical inference under order restrictions* by R. E. Barlow, D. J. Bartholomew, J. M. Bremner, and H. D. Brunk; Van Nostrand Reinhold Co. for permission to quote data from *Smoking and Health*; the Institute of Psychiatry of the University of London for permission to quote data from *Psychiatric diagnosis in New York and London* by J. E. Cooper et al; and Sir Austin Bradford Hill and Oxford University Press for permission to quote from *Statistical methods in clinical and preventive medicine.*

I also thank the editors of the following journals for permission to use published data: the *American Journal of Public Health,* the *American Statistician, Biometrics,* the *Journal of Laboratory and Clinical Medicine,* the *Journal of the National Cancer Institute,* the *Journal of Psychiatric Research* and *Psychometrika.*

JOSEPH L. FLEISS

New York, New York
June 1972

Contents

CHAPTER 1

A Survey of Probability Theory

Some elements of probability theory are needed to appreciate fully and to manipulate the different kinds of rates that arise in research. Thus the clearest and most suggestive interpretation of a rate is as a probability—as a measure of the likelihood that a specified event occurs to, or that a specified characteristic is possessed by, a typical member of a population.

Section 1.1 presents notation and some important definitions. The theory of Section 1.1 is applied in Section 1.2 to the evaluation of a screening test, and in Section 1.3 to the study of the bias possible in making inferences from selected samples.

1.1. NOTATION AND DEFINITIONS

In this book, the terms probability, relative frequency, proportion, and rate are used synonymously. If A denotes the event that a randomly selected individual from a population has a defined characteristic (e.g., has arteriosclerotic heart disease), then $P(A)$ denotes the proportion of all people who have the characteristic. For the given example $P(A)$ is the probability that a randomly selected individual has arteriosclerotic heart disease, or, in the terminology of vital statistics, the case rate for arteriosclerotic heart disease.

One can go only so far with overall rates, however. Of greater usefulness usually are so-called *specific rates:* the rate of the defined characteristic specific for age, race, sex, occupation, and so on. What is known in epidemiology and vital statistics as a specific rate is known in probability theory as a *conditional probability.* The notation is

$P(A \mid B)$ = probability that a randomly selected individual has characteristic A, given that he has characteristic B, or *conditional* on his having characteristic B.

1

2   A SURVEY OF PROBABILITY THEORY

If, in our example, we denote by B the characteristic of being aged 65–74, then $P(A \mid B)$ is the conditional probability that a person has arteriosclerotic heart disease, given that he is aged 65–74. In the terminology of vital statistics, $P(A \mid B)$ is the rate of arteriosclerotic heart disease specific to people aged 65–74.

Let $P(B)$ represent the proportion of all people who possess characteristic B, and let $P(AB)$ represent the proportion of all people who possess both characteristic A and characteristic B. Then, by definition,

$$P(A \mid B) = \frac{P(AB)}{P(B)} . \tag{1.1}$$

Similarly,

$$P(B \mid A) = \frac{P(AB)}{P(A)} . \tag{1.2}$$

By the *association* of two characteristics we mean that, when a person has one of the characteristics, say B, his chances of having the other are affected. By the *independence* or lack of association of two characteristics we mean that the fact that a person has one of the characteristics does not affect his chances of having the other. Thus, if A and B are independent, then the rate at which A is present specific to people who possess B, $P(A \mid B)$, is equal to the overall rate at which A is present, $P(A)$. By (1.1), this implies that

$$\frac{P(AB)}{P(B)} = P(A),$$

or

$$P(AB) = P(A)P(B). \tag{1.3}$$

Equation 1.3 is often taken as the definition of independence, instead of the equivalent statement

$$P(A \mid B) = P(A).$$

A heuristic justification of (1.1) is the following. Let N denote the total number of people in the population; N_A the number of people who have characteristic A; N_B the number of people who have characteristic B; and N_{AB} the number of people who have both characteristics. It is then clear that

$$P(A) = \frac{N_A}{N} ,$$

$$P(B) = \frac{N_B}{N} ,$$

and

$$P(AB) = \frac{N_{AB}}{N} .$$

By $P(A \mid B)$ we mean the proportion, out of all people who have characteristic B, who also have characteristic A, so that both the numerator and the denominator of $P(A \mid B)$ must be specific to B. Thus

$$P(A \mid B) = \frac{N_{AB}}{N_B}. \tag{1.4}$$

If we now divide the numerator and denominator of (1.4) by N, we find that

$$P(A \mid B) = \frac{N_{AB}/N}{N_B/N} = \frac{P(AB)}{P(B)}.$$

Equation 1.2 may be derived similarly:

$$P(B \mid A) = \frac{N_{AB}}{N_A} = \frac{N_{AB}/N}{N_A/N} = \frac{P(AB)}{P(A)}.$$

Equations 1.1 and 1.2 are connected by means of *Bayes' theorem:*

$$P(B \mid A) = \frac{P(A \mid B)P(B)}{P(A)}. \tag{1.5}$$

Equation 1.5 follows from the definition (1.2) of $P(B \mid A)$ and from the fact, seen by multiplying both sides of (1.1) by $P(B)$, that $P(AB) = P(A \mid B)P(B)$.

1.2. THE EVALUATION OF A SCREENING TEST

A frequent application of Bayes' theorem is in evaluating the performance of a diagnostic test intended for use in a screening program. Let B denote the event that a person has the disease in question; $\bar{B}$ the event that he does not have the disease; A the event that he gives a positive response to the test; and $\bar{A}$ the event that he gives a negative response. Suppose that the test has been applied to a sample of B's, that is, to a sample of people with the disease, and to a sample of $\bar{B}$'s, that is, to a sample of people without the disease.

The results of this trial of the screening test may be represented by the two conditional probabilities $P(A \mid B)$ and $P(A \mid \bar{B})$. $P(A \mid B)$ is the conditional probability of a positive response given that the person has the disease; the larger is $P(A \mid B)$, the more *sensitive* is the test. $P(A \mid \bar{B})$ is the conditional probability of a positive response given that the person is free of the disease; the smaller is $P(A \mid \bar{B})$, the more *specific* is the test. These definitions of a test's sensitivity and specificity are due to Yerushalmy (1947).

Of greater concern than the test's sensitivity and specificity, however, are the error rates to be expected if the test is actually used in a screening program.

If a positive result is taken to indicate the presence of the disease, then the false positive rate, say P_{F+}, is the proportion of people, among those responding positive, who are actually free of the disease, or $P(\bar{B} \mid A)$. By Bayes' theorem,

$$P_{F+} = P(\bar{B} \mid A) = \frac{P(A \mid \bar{B})P(\bar{B})}{P(A)} = \frac{P(A \mid \bar{B})(1 - P(B))}{P(A)}, \qquad (1.6)$$

since $P(\bar{B}) = 1 - P(B)$.

The false negative rate, say P_{F-}, is the proportion of people, among those responding negative on the test, who nevertheless have the disease, or $P(B \mid \bar{A})$. Again by Bayes' theorem,

$$P_{F-} = P(B \mid \bar{A}) = \frac{P(\bar{A} \mid B)P(B)}{P(\bar{A})} = \frac{(1 - P(A \mid B))P(B)}{1 - P(A)}, \qquad (1.7)$$

since $P(\bar{A} \mid B) = 1 - P(A \mid B)$ and $P(\bar{A}) = 1 - P(A)$.

We still need the overall rates $P(A)$ and $P(B)$ in order to evaluate these two error rates. Actually, we only need $P(B)$, for the following reason. Note that

$$P(A) = \frac{N_A}{N} = \frac{N_{AB} + N_{A\bar{B}}}{N} = \frac{N_{AB}}{N} + \frac{N_{A\bar{B}}}{N}. \qquad (1.8)$$

In (1.8), N_{AB} denotes the number of people who have the disease and respond positive, and $N_{A\bar{B}}$ denotes the number of people who are free of the disease and respond positive. Multiplying and dividing the first of the two terms on the right-hand side of (1.8) by N_B, the number of people with the disease, we find that

$$\frac{N_{AB}}{N} = \frac{N_{AB}}{N_B} \frac{N_B}{N} = P(A \mid B)P(B). \qquad (1.9)$$

Similarly, by multiplying and dividing the second term by $N_{\bar{B}}$, the number of people without the disease, we find that

$$\frac{N_{A\bar{B}}}{N} = \frac{N_{A\bar{B}}}{N_{\bar{B}}} \frac{N_{\bar{B}}}{N} = P(A \mid \bar{B})P(\bar{B}). \qquad (1.10)$$

Substituting the expressions from (1.9) and (1.10) in (1.8), we find that

$$P(A) = P(A \mid B)P(B) + P(A \mid \bar{B})P(\bar{B}). \qquad (1.11)$$

This equation is a special case of the familiar result that an overall rate—$P(A)$—is a weighted average of specific rates—$P(A \mid B)$ and $P(A \mid \bar{B})$—with the weights being the proportions of people in the specific categories—$P(B)$ and $P(\bar{B})$.

Since $P(\bar{B}) = 1 - P(B)$, (1.11) becomes

$$\begin{aligned} P(A) &= P(A \mid B)P(B) + P(A \mid \bar{B})(1 - P(B)) \\ &= P(A \mid \bar{B}) + P(B)(P(A \mid B) - P(A \mid \bar{B})). \end{aligned} \qquad (1.12)$$

Substitution of (1.12) in (1.6) yields, as the expression for the false positive rate,

$$P_{F+} = \frac{P(A \mid \bar{B})(1 - P(B))}{P(A \mid \bar{B}) + P(B)(P(A \mid B) - P(A \mid \bar{B}))}. \tag{1.13}$$

Substitution of (1.12) in (1.7) yields, as the expression for the false negative rate,

$$P_{F-} = \frac{(1 - P(A \mid B))P(B)}{1 - P(A \mid \bar{B}) - P(B)(P(A \mid B) - P(A \mid \bar{B}))}. \tag{1.14}$$

Analyses of the false positive and false negative rates associated with screening tests have been performed by Meehl and Rosen (1955) and by Dawes (1962) in psychology, by Vastola and Kokubu (1962) in ophthalmology, and by Cochrane and Holland (1971) for a variety of medical disorders. We see from (1.13) and (1.14) that, in general, the two error rates are functions of the proportions $P(A \mid B)$ and $P(A \mid \bar{B})$, which may be estimated from the results of a trial of the screening test, and of the overall case rate $P(B)$, for which an accurate estimate is rarely available. Nevertheless, a range of likely values for the error rates may be determined as in the following example.

Suppose that the test is applied to a sample of 1000 people known to have the disease and to a sample of 1000 people known not to have the disease. Suppose that this trial resulted in the frequencies shown in Table 1.1. We

Table 1.1. Results of a trial of a screening test

Disease Status	Test Result		Total
	$+(A)$	$-(\bar{A})$	
Present (B)	950	50	1000
Absent $(\bar{B})$	10	990	1000

would then have the estimates

$$P(A \mid B) = 950/1000 = .95$$

and

$$P(A \mid \bar{B}) = 10/1000 = .01,$$

a pair of probabilities indicating a test that is sensitive [$P(A \mid B)$ is close to unity] and specific [$P(A \mid \bar{B})$ is close to zero] to the disease being studied.

Substitution of these two probabilities in (1.13) gives, as the value for the false positive rate,

$$P_{F+} = \frac{.01(1 - P(B))}{.01 + P(B)(.95 - .01)} = \frac{.01(1 - P(B))}{.01 + .94P(B)}$$

$$= \frac{1 - P(B)}{1 + 94P(B)},$$ (1.15)

the final expression resulting by multiplying both the numerator and the denominator of the preceding expression by 100. Substitution in (1.14) gives, as the value for the false negative rate,

$$P_{F-} = \frac{(1 - .95)P(B)}{1 - .01 - P(B)(.95 - .01)} = \frac{.05P(B)}{.99 - .94P(B)}$$

$$= \frac{5P(B)}{99 - 94P(B)}.$$ (1.16)

Table 1.2 gives the error rates associated with various values of $P(B)$, the overall case rate. Rarely will the case rate exceed 1 % of the population.

Table 1.2. Error rates associated with screening test

$P(B)$	False Positive (P_{F+})	False Negative (P_{F-})
1/million	.9999	0
1/100,000	.9991	0
1/10,000	.9906	.00001
1/1000	.913	.00005
1/500	.840	.00010
1/200	.677	.00025
1/100	.510	.00051

If the disease is not too prevalent—if it affects, say, less than 1 % of the population—the false negative rate will be quite small, but the false positive rate will be rather large. From one point of view, the test is a successful one: since P_{F-} is less than 5/10,000, therefore, of every 10,000 people who respond negative and are thus presumably given a clean bill of health, no more than five should actually have been informed they were ill. From another point of view, the test is a failure: since P_{F+} is greater than 50/100, therefore, of every 100 people who respond positive and thus presumably are told they have the disease or are at least advised to undergo further tests, more than 50 will actually be free of the disease.

The final decision whether or not to use the test will depend on the seriousness of the disease and on the cost of further tests or treatment. Because the false positive rate is so great, however, it would be hard to justify using this screening test for any but the most serious diseases. Since the proportions $P(A \mid B)$ and $P(A \mid \bar{B})$ considered in this example are better than those associated with most existing screening tests, it is disquieting to realize that their false positive rates are probably greatly in excess of the 50 out of 100 that we found as a lower limit here.

One method for reducing the false positive or false negative rate associated with a diagnostic screening procedure (but thereby increasing its cost) is to repeat the test a number of times, and to declare the final result positive if the subject responds positive to each administration of the test or if he responds positive to a majority of the administrations. Parts b and c of Problem 1.2 illustrate the improved performance of a screening procedure when a test is administered twice. Sandifer, Fleiss, and Green (1968) have shown that, for some disorders, a better rule is to administer the test three times and to declare the final result positive if the subject responds positive to at least two of the three administrations. Only those subjects who respond positive to one of the first two administrations and negative to the other will have to be tested a third time. Those who respond positive to both of the first two administrations would be declared positive, and those who respond negative to both would be declared negative.

A more accurate but more complex assessment of the performance of a screening procedure than the above is possible when disease severity is assumed to vary and not, as here, merely to be present or absent. The appropriate analysis was originally proposed by Neyman (1947) and later extended by Greenhouse and Mantel (1950) and by Nissen-Meyer (1964). The reader is referred to these papers for details.

1.3. BIASES RESULTING FROM THE STUDY OF SELECTED SAMPLES

The first clues to the association between diseases and antecedent conditions frequently come from the study of such selected samples as hospitalized patients and autopsy cases. Because not all subjects are equally likely to end up in the study samples, bias may result when the associations found in the selected samples are presumed to apply in the population at large.

A classic example of this kind of bias occurs in a study by Pearl (1929). A large number of autopsy cases were cross-classified by the presence or absence of cancer and by the presence or absence of tuberculosis. A negative association between these two diseases was found, that is, tuberculosis was less frequent in autopsy cases with cancer than in cases without cancer. Pearl inferred that the same negative association should apply to live patients, and in fact acted on the basis of this inference by conceiving a study to treat terminal cancer patients with tuberculin (the protein of the tubercle bacillus) in the anticipation that the cancer would be arrested. What Pearl ignored is that, unless all deaths are equally likely to be autopsied, it is improper to extrapolate to live patients an association found for autopsied cases. It is even possible that there would be no association for live patients but, due to the differential selection of patients for autopsy, a strong association for autopsied cases.

The same kind of bias is possible whenever the chances of a subject's being included in the study sample vary. This has been pointed out by Berkson (1946, 1955), Mainland (1953, 1963, pp. 117–124), White (1953), and Mantel and Haenszel (1959). The bias is illustrated using hypothetical data.

Suppose that a research worker in a general hospital, reviewing data for a number of patients, comes up with the fourfold table shown as Table 1.3.

Table 1.3. Prior living arrangement by diagnosis

	Prior Living Arrangement			
Diagnosis	Alone	With Family	Total	Proportion Alone
Neurotic	48	52	100	$.48 = p_1$
Non–neurotic	104	696	800	$.13 = p_2$
Total	152	748	900	

There is clearly an association between whether a patient is or is not neurotic and whether he did or did not live alone: the proportion of neurotics who had lived alone, $p_1 = .48$, is nearly four times the proportion of non-neurotics who lived alone, $p_2 = .13$. Would it, however, be correct to conclude that type of living arrangement and neuroticism were associated in the community?

Not necessarily. These two characteristics—neuroticism and living arrangement—may be independent in the community and yet end up as associated in the hospital. This phenomenon is always possible when admission rates for people with different combinations of factors vary, and can really be ruled out only when the disease in question almost always requires care (e.g., leukemia and other cancers).

The algebra underlying the phenomenon is as follows. Let B denote the event that a person has the disease (in this case, a neurosis) and $\bar{B}$ the event that a person does not have the disease. Let $P(B)$ denote the proportion of all people in the community who have the disease and let $P(\bar{B}) = 1 - P(B)$ denote the proportion of all people who are free of the disease.

Let A denote the event that a person lives alone and $\bar{A}$ the event that a person lives with his family, and let $P(A)$ and $P(\bar{A}) = 1 - P(A)$ represent the corresponding proportions. Let $P(A$ and $B)$ denote the proportion of all people in the community who *both* have the disease and live alone, and assume that these two characteristics are independent in the community. Thus, by (1.3), $P(A$ and $B) = P(A)P(B)$.

Let H denote the event that a person from the community is hospitalized for some reason or other. Define

$P(H \mid B$ and $A) = $ the proportion, out of all people who have the disease and who live alone, who are hospitalized;

$P(H \mid B$ and $\bar{A}) = $ the proportion, out of all people who have the disease and who live with their families, who are hospitalized;

and define $P(H \mid \bar{B}$ and $A)$ and $P(H \mid \bar{B}$ and $\bar{A})$ similarly. Our problem is to evaluate, in terms of these probabilities,

$$p_1 = P(A \mid B \text{ and } H),$$

that is, the proportion, out of all people who have the disease and are hospitalized, who live alone; and

$$p_2 = P(A \mid \bar{B} \text{ and } H),$$

that is, the proportion, out of all people who do not have the disease and are hospitalized, who live alone.

We make use of the following version of the definition of a conditional probability:

$$p_1 = P(A \mid B \text{ and } H) = \frac{P(B \text{ and } A \mid H)}{P(B \mid H)}. \tag{1.17}$$

Equation 1.17 differs from (1.2) only in that, once specified, the second condition H (for hospitalization) remains a condition qualifying all probabilities.

The numerator of (1.17) is, by Bayes' theorem [see (1.5)],

$$P(B \text{ and } A \mid H) = \frac{P(H \mid B \text{ and } A)P(B \text{ and } A)}{P(H)}$$

$$= \frac{P(H \mid B \text{ and } A)P(B)P(A)}{P(H)}, \tag{1.18}$$

because of the assumed independence of A and B.

The denominator of (1.17) is, again by Bayes' theorem,

$$P(B \mid H) = \frac{P(H \mid B)P(B)}{P(H)}. \tag{1.19}$$

To find $P(H \mid B)$, we make use of the fact, represented by (1.11), that an overall rate is a weighted average of specific rates. This time, however, we apply the additional fact that a condition on a probability, in this case B, remains one in all subsequent probabilities. Thus

$$P(H \mid B) = P(H \mid B \text{ and } A)P(A \mid B) + P(H \mid B \text{ and } \bar{A})P(\bar{A} \mid B)$$
$$= P(H \mid B \text{ and } A)P(A) + P(H \mid B \text{ and } \bar{A})P(\bar{A}),$$

because the assumed independence of A and B implies that $P(A \mid B) = P(A)$ and $P(\bar{A} \mid B) = P(\bar{A})$. Therefore, (1.19) becomes

$$P(B \mid H) = \frac{P(B)[P(H \mid B \text{ and } A)P(A) + P(H \mid B \text{ and } \bar{A})P(\bar{A})]}{P(H)}. \quad (1.20)$$

Substitution of (1.18) for the numerator of (1.17), and of (1.20) for the denominator, yields

$$p_1 = \frac{P(H \mid B \text{ and } A)P(A)}{P(H \mid B \text{ and } A)P(A) + P(H \mid B \text{ and } \bar{A})P(\bar{A})}. \quad (1.21)$$

Similarly,

$$p_2 = \frac{P(H \mid \bar{B} \text{ and } A)P(A)}{P(H \mid \bar{B} \text{ and } A)P(A) + P(H \mid \bar{B} \text{ and } \bar{A})P(\bar{A})}. \quad (1.22)$$

These two probabilities are not equal if the rates of hospitalization are not equal, even though the corresponding probabilities in the community, $P(A \mid B)$ and $P(A \mid \bar{B})$, are equal.

As an example, suppose that 60% of all people live alone, so that $P(A) = .60$, and suppose that the various hospitalization rates are

$$P(H \mid B \text{ and } A) = .25,$$
$$P(H \mid B \text{ and } \bar{A}) = .40,$$
$$P(H \mid \bar{B} \text{ and } A) = .01,$$

and

$$P(H \mid \bar{B} \text{ and } \bar{A}) = .10.$$

These rates are such that neurotics, whether or not they live with their families, have higher hospitalization rates than non-neurotics, and that people who live with their families, both neurotics and non-neurotics, have higher hospitalization rates than people who live alone.

Substituting these values into (1.21) and (1.22), we find that

$$p_1 = \frac{.25 \times .60}{.25 \times .60 + .40 \times .40} = \frac{.15}{.31} = .48$$

and that

$$p_2 = \frac{.01 \times .60}{.01 \times .60 + .10 \times .40} = \frac{.006}{.046} = .13,$$

precisely the values we found in Table 1.3.

The moral of this exercise is clear. Unless something is known about differential hospitalization rates or differential autopsy rates, a good amount of skepticism should be applied to any generalization from associations found for hospitalized patients or for autopsy cases to associations for people at large. This caveat obviously applies also to associations obtained from reports by volunteers.

Problem 1.1. Characteristics A and B are independent if $P(A \text{ and } B) = P(A)P(B)$. Show that, if this is so, then $P(A \text{ and } \bar{B}) = P(A)P(\bar{B})$; $P(\bar{A} \text{ and } B) = P(\bar{A})P(B)$; and $P(\bar{A} \text{ and } \bar{B}) = P(\bar{A})P(\bar{B})$. [*Hint.* $P(A) = P(A \text{ and } B) + P(A \text{ and } \bar{B})$, so that $P(A \text{ and } \bar{B}) = P(A) - P(A \text{ and } B)$. Use the given relation, $P(A \text{ and } B) = P(A)P(B)$, and use the fact that $P(\bar{B}) = 1 - P(B)$.]

Problem 1.2. Assume that the case rate for a specified disease, $P(B)$, is one case per 1000 population and that a screening test for the disease is being studied.

(*a*) Suppose that the test is administered a single time to a sample of people with the disease, of whom 99% respond positive. Suppose that it is also administered to a sample of people without the disease, of whom 1% respond positive. What are the false positive and false negative rates? Do you think the test is a good one?

(*b*) Suppose now that the test is administered twice, with a positive overall result declared only if both tests are positive. Suppose, according to this revised definition, that 95% of the diseased sample respond positive, but that only one out of 10,000 nondiseased people responds positive. What are the false positive and false negative rates now? Would you be willing to employ the test for screening under these revised conditions?

(*c*) Note that not all people have to be tested twice. If the first test result is negative, the person does not have to be tested again. Only if the first test result is positive must a second test be administered; the final result will be declared positive only if the second test is positive as well. It is important to estimate the proportion of all people who will have to be tested again, that is, the proportion who are positive on their first test. What is this proportion? Out of every 100,000 people tested once, how many will not need to be tested again?

Problem 1.3. The opposite kind of bias from that considered in Section 1.3 can occur, that is, two characteristics may be associated in the community but may be independent when hospitalized samples are studied. Let the events A and B represent what they did above, and suppose that $P(A \mid B) = .40$ and $P(A \mid \bar{B}) = .20$. Thus 40% of neurotics live alone and 20% of non-neurotics live alone.

Suppose that, in the population at large, 100,000 people are neurotic and one million are not neurotic.

(*a*) Consider first the 100,000 neurotics. (1) How many of them live alone? (2) If the annual hospitalization rate for neurotics living alone is 5/1000, how many such people will be hospitalized? Note that this is the number of hospitalized neurotics one would find who lived alone, that is, the numerator of p_1. (3) How many of the 100,000 neurotics live with their families? (4) If the annual hospitalization rate for neurotics living with their families is 6/1000, how many such people will be hospitalized? Note that the sum of the numbers found in (2) and (4) is the total number of hospitalized neurotics, that is, the denominator of p_1. (5) What is the value of p_1, the proportion of hospitalized neurotics who lived alone? (6) How does p_1 compare with $P(A \mid B)$, the proportion of neurotics in the community who live alone?

(*b*) Consider now the one million non-neurotics. (1) How many of them live alone? (2) If the annual hospitalization rate for non-neurotics living alone is 5/1000, how many such people will be hospitalized? Note that this is the number of hospitalized non-neurotics one would find who lived alone, that is, the numerator of p_2. (3) How many of the one million non-neurotics live with their families? (4) If the annual hospitalization rate for non-neurotics living with their families is 225/100,000, how many such people will be hospitalized? Note that the sum of the numbers found in (*b*2) and (*b*4) is the total number of hospitalized non-neurotics, that is, the denominator of p_2. (5) What is the value of p_2, the proportion of hospitalized non-neurotics who lived alone? (6) How does p_2 compare with $P(A \mid \bar{B})$, the proportion of non-neurotics in the community who live alone?

(*c*) What inference would you draw from the comparison of p_1 and p_2? How does this inference compare with the inference drawn from the comparison of $P(A \mid B)$ and $P(A \mid \bar{B})$?

REFERENCES

Berkson, J. (1946). Limitations of the application of fourfold table analysis to hospital data. *Biometrics Bull.* (now *Biometrics*), **2**, 47–53.

Berkson, J. (1955). The statistical study of association between smoking and lung cancer. *Proc. Staff Meetings Mayo Clin.*, **30**, 319–348.

Cochrane, A. L. and Holland, W. W. (1971). Validation of screening procedures. *Brit. med. Bull.*, **27**, 3–8.

Dawes, R. M. (1962). A note on base rates and psychometric efficiency. *J. consult. Psychol.*, **26**, 422–424.

Greenhouse, S. W. and Mantel, N. (1950). The evaluation of diagnostic tests. *Biometrics*, **6**, 399–412.

Mainland, D. (1953). The risk of fallacious conclusions from autopsy data on the incidence of diseases with applications to heart disease. *Amer. Heart J.*, **45**, 644–654.

Mainland, D. (1963). *Elementary medical statistics, 2nd ed.* Philadelphia: W. W. Saunders.

Mantel, N. and Haenszel, W. (1959). Statistical aspects of the analysis of data from retrospective studies of disease. *J. natl. Cancer Inst.*, **22**, 719–748.

Meehl, P. E. and Rosen, A. (1955). Antecedent probability and the efficiency of psychometric signs, patterns, or cutting scores. *Psychol. Bull.*, **52**, 194–216.

Neyman, J. (1947). Outline of statistical treatment of the problem of diagnosis. *Public Health Rep.*, **62**, 1449–1456.

Nissen-Meyer, S. (1964). Evaluation of screening tests in medical diagnosis. *Biometrics*, **20**, 730–755.

Pearl, R. (1929). Cancer and tuberculosis. *Amer. J. Hyg.* (now *Amer. J. Epidemiol.*), **9**, 97–159.

Sandifer, M. G., Fleiss, J. L. and Green, L. M. (1968). Sample selection by diagnosis in clinical drug evaluations. *Psychopharm.*, **13**, 118–128.

Vastola, E. F. and Kokubu, T. (1962). Use of Bayes' theorem in ophthalmodynamometry. *Circulation*, **26**, 1312–1315.

White, C. (1953). Sampling in medical research. *Brit. med. J.*, **2**, 1284–1288.

Yerushalmy, J. (1947). Statistical problems in assessing methods of medical diagnosis, with special reference to X-ray techniques. *Public Health Rep.*, **62**, 1432–1449.

CHAPTER 2

Assessing Significance in a Fourfold Table

The fourfold table (see Table 2.1) has been and probably still is the most frequently employed means of presenting statistical evidence. The simplest

Table 2.1. **Model fourfold table**

| | Characteristic B | | |
Characteristic A	Present	Absent	Total
Present	n_{11}	n_{12}	$n_{1.}$
Absent	n_{21}	n_{22}	$n_{2.}$
Total	$n_{.1}$	$n_{.2}$	$n_{..}$

and most frequently applied statistical test of the significance of the association indicated by the data is the classical chi square test. It is based on the magnitude of the statistic

$$\chi^2 = \frac{n_{..}(|n_{11}n_{22} - n_{12}n_{21}| - \tfrac{1}{2}n_{..})^2}{n_{1.}n_{2.}n_{.1}n_{.2}}. \tag{2.1}$$

The value obtained for χ^2 is referred to tables of the chi square distribution with one degree of freedom (see Table A.1). If the value exceeds the entry tabulated for a specified significance level, the inference is made that A and B are associated. An interesting graphic assessment of significance is due to Zubin (1939).

Table 2.2 presents some hypothetical frequencies. The value of χ^2 is, by (2.1),

$$\chi^2 = \frac{200(|15 \times 40 - 135 \times 10| - \tfrac{1}{2}200)^2}{150 \times 50 \times 25 \times 175} = 2.58. \tag{2.2}$$

Since χ^2 would have to exceed 3.84 in order for significance at the .05 level

Table 2.2. A hypothetical fourfold table

Characteristic A	Characteristic B		Total
	Present	Absent	
Present	15	135	150
Absent	10	40	50
Total	25	175	200

to be declared, the conclusion for these data would be that no significant association was demonstrated.

It is perhaps unfortunate that the chi square statistic (2.1) takes such a simple form, both because its calculation does not require the investigator to determine explicitly the proportions being contrasted—these representing the association being studied, not the raw frequencies—and because it invites the investigator to ignore the fact that the proper inference to be drawn from the magnitude of χ^2 depends on how the data were generated, even though the formula for χ^2 does not. These ideas are developed in Section 2.1. There, three methods for generating the frequencies of a fourfold table are presented and the statistical hypothesis appropriate to each is specified.

In Section 2.2 the need for applying Yates' correction for continuity is considered. In Section 2.3 some criteria for choosing between a one-tailed and a two-tailed significance test are offered.

2.1. METHODS FOR GENERATING A FOURFOLD TABLE

There are three basic methods of sampling that can give rise to the frequencies set out in a fourfold table.

Method I. The first method of sampling, termed naturalistic or cross-sectional sampling, calls for the selection of a total of $n_{..}$ subjects from a larger population, and then for the determination for each subject of the presence or absence of characteristic A and the presence or absence of characteristic B. Only the total sample size, $n_{..}$, can be specified prior to the collection of the data.

Much of survey research is conducted along such a line. Examples of the use of method I sampling are the following. In a study of the quality of medical care delivered to patients, all new admissions to a specified service of a hospital might be cross-classified by sex and by whether or not each of a number of examinations was made. In a study of the variation of disease prevalence in a community, a random sample of subjects may be drawn and cross-classified by race and by the presence or absence of each of a number of symptoms. In a study of the association between birthweight and maternal age, all deliveries in a given maternity hospital might be cross-classified by the weight of the offspring and by the age of the mother.

With method I sampling, the issue is whether the presence or absence of characteristic A is associated with the presence or absence of characteristic B. In the population from which the sample was drawn, the proportions (of course unknown) are as in Table 2.3.

Table 2.3. Joint proportions of A and B in the population

	Characteristic B		
Characteristic A	Present	Absent	Total
Present	P_{11}	P_{12}	$P_{1.}$
Absent	P_{21}	P_{22}	$P_{2.}$
Total	$P_{.1}$	$P_{.2}$	1

By the definition of independence (see Section 1.1), characteristics A and B are independent if and only if each joint proportion (e.g., P_{12}) is the product of the two corresponding total or marginal proportions (in this example, $P_{1.}P_{.2}$). Whether the proportions actually have this property can only be determined by how close the joint proportions in the sample are to the corresponding products of marginal proportions. The cross-classification table in the sample should therefore be the analog of Table 2.3, and is obtained by dividing each frequency in Table 2.1 by $n_{..}$. Table 2.4 results.

Table 2.4. Joint proportions of A and B in the sample

	Characteristic B		
Characteristic A	Present	Absent	Total
Present	p_{11}	p_{12}	$p_{1.}$
Absent	p_{21}	p_{22}	$p_{2.}$
Total	$p_{.1}$	$p_{.2}$	1

The tenability of the hypothesis that A and B are independent depends on the magnitudes of the four differences $p_{ij} - p_{i.}p_{.j}$, where i and j equal 1 or 2. The smaller these differences are, the closer do the data come to the standard of independence. The larger these differences are, the more questionable the hypothesis of independence becomes. (Actually, only a single one of these four differences needs to be examined, the other three being equal to it except possibly for a change in sign—see Problem 2.1.)

Pearson (1900) suggested a criterion for assessing the significance of these differences. His statistic, incorporating the continuity correction, is

$$\chi^2 = n_{..} \sum_{i=1}^{2} \sum_{j=1}^{2} \frac{(|p_{ij} - p_{i.}p_{.j}| - 1/(2n_{..}))^2}{p_{i.}p_{.j}}. \tag{2.3}$$

Problem 2.2 is devoted to the proof that (2.1) and (2.3) are equal. If χ^2 is found by reference to the table of chi square with one degree of freedom to be significantly large, the investigator would infer that A and B were associated and would proceed to describe their degree of association. Chapter 5 is devoted to methods of describing association following method I sampling.

Suppose that the data of Table 2.2 were obtained from a study employing method I sampling. The proper summarization of the data is illustrated by Table 2.5. Note, for example, that $p_{22} = .20$, whereas if A and B were

Table 2.5. Joint proportions for hypothetical data of Table 2.2

	Characteristic B		
Characteristic A	Present	Absent	Total
Present	.075	.675	.75
Absent	.05	.20	.25
Total	.125	.875	1

independent, we would have expected the proportion to be $p_{2.}p_{.2} = .25 \times .875 = .21875$. Furthermore, note that each of the four differences entering into (2.3) is equal to $|\pm .01875| - .0025 = .01625$.

Applied to the data of Table 2.5, (2.3) yields the value

$$\chi^2 = 200\left(\frac{.01625^2}{.09375} + \frac{.01625^2}{.65625} + \frac{.01625^2}{.03125} + \frac{.01625^2}{.21875}\right)$$
$$= 2.58, \tag{2.4}$$

equal to the value in (2.2).

Method II. The second method of sampling, sometimes termed purposive sampling, calls for the selection and study of a predetermined number, $n_{1.}$, of subjects who possess characteristic A, and for the selection and study of a predetermined number, $n_{2.}$, of subjects for whom characteristic A is absent. This method of sampling forms the basis of comparative prospective and of comparative retrospective studies. In the former, $n_{1.}$ subjects with and $n_{2.}$ subjects without a suspected antecedent factor would be followed to determine how many develop disease. In the latter, $n_{1.}$ subjects with and $n_{2.}$

subjects without the disease would be traced back to determine how many possessed the suspected antecedent factor.

Of interest in method II sampling is whether the proportions in the two populations from which we have samples, say P_1 and P_2, are equal. It is therefore indicated that the sample data be so presented that information about these two proportions is afforded. The appropriate means of presentation is given in Table 2.6.

Table 2.6. Proportions with a specified characteristic in two independent samples

	Sample Size	Proportion
Sample 1	$n_{1.}$	$p_1 \ (= n_{11}/n_{1.})$
Sample 2	$n_{2.}$	$p_2 \ (= n_{21}/n_{2.})$
Combined	$n_{..}$	$\bar{p} \ (= n_{.1}/n_{..})$

The statistical significance of the difference between p_1 and p_2 is assessed by means of the statistic

$$z = \frac{|p_2 - p_1| - \frac{1}{2}(1/n_{1.} + 1/n_{2.})}{\sqrt{\bar{p}\bar{q}(1/n_{1.} + 1/n_{2.})}}, \qquad (2.5)$$

where $\bar{q} = 1 - \bar{p}$. If P_1 and P_2 are equal, then z should have the standard normal distribution. If z exceeds the normal curve value for a prespecified significance level—see Table A.2—P_1 and P_2 are inferred to be unequal. Since, by definition, the square of a quantity that has the standard normal distribution will be distributed as chi square with one degree of freedom, therefore z^2 may be referred to tables of chi square with one degree of freedom. Problem 2.3 is devoted to the proof that z^2 is equal to the quantity in (2.1).

The analysis subsequent to the finding of statistical significance with method II sampling is shown in Chapter 6 to be quite different from the analysis appropriate to method I sampling.

Suppose, for illustration, that the data of Table 2.2 had been generated by deliberately studying 150 subjects with characteristic A and 50 subjects without it. The appropriate presentation of the data is illustrated in Table 2.7.

Table 2.7. Proportions with characteristic B for subjects with and subjects without characteristic A—from hypothetical data of Table 2.2

	Sample Size	Proportion with B
A Present	150	$.10 = p_1$
A Absent	50	$.20 = p_2$
Total	200	$.125 = \bar{p}$

The value of z (2.5) is

$$z = \frac{|.20 - .10| - \frac{1}{2}(\frac{1}{150} + \frac{1}{50})}{\sqrt{.125 \times .875(\frac{1}{150} + \frac{1}{50})}} = 1.60, \qquad (2.6)$$

which fails to reach the value 1.96 needed for significance at the .05 level. The square of the obtained value of z is 2.56, equal except for rounding errors to the value of χ^2 in (2.2).

Method III. The third method of sampling is like method II in that two samples of predetermined size are contrasted. Unlike method II, however, method III calls for the two samples to be constituted at random. This method lies at the basis of the controlled comparative clinical trial: of a total of $n_{..}$ subjects, $n_{1.}$ are selected at random to be treated with the control treatment and the remaining $n_{2.}$ to be treated with the test treatment.

Of importance are the proportions from the two groups experiencing the outcome under study (e.g., the remission of symptoms). The significance of their difference is assessed by the same statistic (2.5) appropriate to method II. The appropriate further description of the data, however, is shown in Chapter 7 to be different for the two methods.

2.2. YATES' CORRECTION FOR CONTINUITY

Yates (1934) suggested that the correction

$$C_1 = -\tfrac{1}{2}n_{..} \qquad (2.7)$$

be incorporated into expression (2.1) for χ^2 and that the correction

$$C_2 = -\frac{1}{2}\left(\frac{1}{n_{1.}} + \frac{1}{n_{2.}}\right) \qquad (2.8)$$

be incorporated into expression (2.5) for z. These corrections take account of the fact that a continuous distribution (the chi square and normal, respectively) is being used to represent the discrete distribution of sample frequencies.

Studies of the effects of the continuity correction have been made by Pearson (1947), Pearson and Hartley (1958, pp. 68–72), Mote, Pavate, and Anderson (1958), and Plackett (1964). On the basis of these and of their own analyses, Grizzle (1967) and Conover (1968) recommend that the correction for continuity not be applied. They give as their reason an apparent lowering of the actual significance level when the correction is used. A lowered significance level results in a reduction in *power*, that is, in a reduced probability of detecting a real association or real difference in rates.

Mantel and Greenhouse (1968) point out the inappropriateness of Grizzle's (and, by implication, of Conover's) analyses, and refute their argument against the use of the correction. The details of Mantel and Greenhouse's refutation are beyond the scope of this book. An outline of their reasoning is provided instead.

In method I, the investigator hypothesizes no association between factors A and B, which means that all four cell probabilities are functions of the marginal proportions $P_{1.}$, $P_{2.}$, $P_{.1}$, and $P_{.2}$ (see Table 2.3). Because the investigator is almost never in the position to specify what the values of these proportions are, he must use his obtained marginal frequencies to estimate them.

In methods II and III, the investigator hypothesizes no difference between two independent proportions, P_1 and P_2. Because the investigator is almost never in the position to specify what the value of the hypothesized common proportion is, he must use his obtained marginal frequencies to estimate it.

For each of the three sampling methods, the investigator must therefore proceed to analyze his data with the restriction that his marginal proportions instead of the unknown population proportions characterize the factors under study. This restriction is equivalent to considering the four marginal frequencies $n_{1.}$, $n_{2.}$, $n_{.1}$, and $n_{.2}$ he obtained (see Table 2.1) as fixed. Under the restriction of fixed marginal frequencies, exact probabilities associated with the cell frequencies n_{11}, n_{12}, n_{21}, and n_{22} may be derived from the *hypergeometric probability distribution*. This distribution is described in most statistics and probability texts, and forms the basis of the exact test due to Fisher (1934) and Irwin (1935). Because the correction for continuity brings probabilities associated with χ^2 and z into close agreement with the exact probabilities, the correction should always be used.

2.3. ONE-TAILED VERSUS TWO-TAILED TESTS

The chi square test and equivalent normal curve test so far considered are examples of *two-tailed* tests. Specifically, a significant difference is declared either if p_2 is sufficiently *greater* than p_1 or if p_2 is sufficiently *less* than p_1. Suppose that the investigator is interested only in a difference in one direction, say in P_2, the underlying proportion in group 2, being greater than P_1, the underlying proportion in group 1. He can increase the power of his comparison by performing a *one-tailed* test. The investigator can make one of two inferences after a one-tailed test, either that p_2 is significantly greater than p_1 or that it is not. By performing a one-tailed test, he is ruling out as unimportant the possible inference that p_1 is significantly greater than p_2.

The one-tailed test begins with an inspection of the data to see if they are in the direction specified by the hypothesis. If they are not (e.g., if $p_1 \geq p_2$ but the investigator was only interested in a difference in the reverse direction), no further calculations are performed and the inference is made that P_2 might not be greater than P_1. If the data are consistent with the hypothesis, the investigator proceeds to calculate either the χ^2 statistic (2.1) or the z statistic (2.5).

The magnitude of χ^2 is assessed for significance as follows. If the investigator desires to have a significance level of α, he enters Table A.1 under the column for 2α. If his calculated value of χ^2 exceeds the tabulated critical value, he infers that the underlying proportions differ in the hypothesized direction (e.g., that $P_2 > P_1$). If not, he infers that the underlying proportions might not differ in the hypothesized direction. The magnitude of z is assessed similarly. When the desired significance level is α, Table A.2 is entered with 2α.

It is seen from Tables A.1 and A.2 that critical values for a significance level of 2α are less than those for a significance level of α. An obtained value for the test statistic (either χ^2 or z) which fails to exceed the critical value for a significance level of α may nevertheless exceed the critical value for a significance level of 2α. Because it is easier to reject a hypothesis of no difference with a one-tailed than with a two-tailed test when the proportions differ in the hypothesized direction, the former test is more powerful than the latter.

As presented here, a one-tailed test is called for only when the investigator is not interested in a difference in the reverse direction from that hypothesized. For example, if he hypothesizes that $P_2 > P_1$, then it will make no difference to him if either $P_2 = P_1$ or $P_2 < P_1$. Such an instance is assuredly rare. One example where a one-tailed test is called for is when an investigator is comparing the response rate for a new treatment (p_2) with the response rate for a standard treatment (p_1), and when he will substitute the new treatment for the standard in his own practice only if p_2 is significantly greater than p_1. It will make no difference to him if the two treatments are equally effective or if the new treatment is actually worse than the standard; in either case, he will stick with the standard.

If, however, the investigator intends to report his results to his professional colleagues, he is ethically bound to perform a two-tailed test. For if his results indicate that the new treatment is actually worse than the standard—an inference possible only with a two-tailed test—he is obliged to report this as a warning to others who might plan to study the new treatment.

In the vast majority of research undertakings, two-tailed tests are called for. Even if a theory or a large accumulation of published data suggests that the difference being studied should be in one direction and not the other, the investigator should nevertheless guard against the unexpected by performing a two-tailed test. Especially in such cases, the scientific importance of a difference in the unexpected direction may be greater than yet another confirmation of the difference being in the expected direction.

Problem 2.1. Consider the joint proportions of Table 2.4. Prove that $p_{12} - p_{1.}p_{.2} = -(p_{11} - p_{1.}p_{.1})$; that $p_{21} - p_{2.}p_{.1} = -(p_{11} - p_{1.}p_{.1})$; and that $p_{22} - p_{2.}p_{.2} = p_{11} - p_{1.}p_{.1}$. (*Hint.* Because $p_{11} + p_{12} = p_{1.}$, therefore $p_{12} = p_{1.} - p_{11}$. Use the fact that $1 - p_{.2} = p_{.1}$.)

Problem 2.2. Prove that formulas (2.3) and (2.1) for χ^2 are equal. (*Hint.* Begin by using the result of Problem 2.1 to factor $(|p_{11} - p_{1.}p_{.1}| - 1/(2n_{..}))^2$ out of the summation in formula (2.3). Bring the four remaining terms, $1/(p_{i.}p_{.j})$, over a common denominator and show, using the facts that $p_{1.} + p_{2.} = p_{.1} + p_{.2} = 1$, that the numerator of the resulting expression is unity. Finally, replace each proportion by its corresponding ratio of frequencies.)

Problem 2.3. Prove that the square of z—see (2.5)—is equal to the expression for χ^2 given in (2.1).

REFERENCES

Conover, W. J. (1968). Uses and abuses of the continuity correction. *Biometrics*, **24**, 1028.

Fisher, R. A. (1934). *Statistical methods for research workers, 5th ed.* Edinburgh: Oliver and Boyd.

Grizzle, J. E. (1967). Continuity correction in the χ^2-test for 2 × 2 tables. *Amer. Statist.*, **21** (October), 28–32.

Irwin, J. O. (1935). Tests of significance for differences between percentages based on small numbers. *Metron*, **12**, 83–94.

Mantel, N. and Greenhouse, S. W. (1968). What is the continuity correction? *Amer. Statist.*, **22** (December), 27–30.

Mote, V. L., Pavate, M. V. and Anderson, R. L. (1958). Some studies in the analysis of categorical data. *Biometrics*, **14**, 572–573.

Pearson, E. S. (1947). The choice of statistical tests illustrated on the interpretation of data classed in a 2 × 2 table. *Biometrika*, **34**, 139–167.

Pearson, E. S. and Hartley, H. O. (Eds.) (1958). *Biometrika tables for statisticians*, Vol. I. Cambridge, England: Cambridge University Press.

Pearson, K. (1900). On the criterion that a given system of deviations from the probable in the case of a correlated system of variables is such that it can be reasonably supposed to have arisen from random sampling. *Philos. Mag.*, **50**(5), 157–175.

Plackett, R. L. (1964). The continuity correction in 2 × 2 tables. *Biometrika*, **51**, 327–337.

Yates, F. (1934). Contingency tables involving small numbers and the χ^2 test. *J. roy. statist. Soc. Suppl.*, **1**, 217–235.

Zubin, J. (1939). Nomographs for determining the significance of the differences between the frequencies of events in two contrasted series or groups. *J. Amer. statist. Assoc.*, **34**, 539–544.

CHAPTER 3

Determining Sample Sizes Needed to Detect a Difference Between Two Proportions

There are two kinds of errors one must guard against in designing a comparative study. Even though these errors can occur in any statistical evaluation, their discussion here is restricted to the case where proportions from two independent samples are compared, that is, to sampling methods II and III. The reader is referred to Cohen (1969, Chapter 7) for a discussion of the two kinds of errors in sampling method I.

The first error, called the Type I error, consists in declaring that the difference in proportions being studied is real when in fact the difference is zero. This error has been given the greater amount of attention in elementary statistics books, and hence in practice. It is typically guarded against simply by setting the significance level for the chosen statistical test, denoted α, at a suitably small probability such as .01 or .05.

This kind of control is not totally adequate, because a Type I error probably never occurs in practice. The reason is that the two populations giving rise to the observed samples will inevitably differ to some extent, albeit possibly by a trivially small amount. This is as true in the case of the improvement rates associated with any two treatments as in the case of the disease rates for people possessing and for people not possessing any suspected antecedent factor. It is shown in Problem 3.2 that no matter how small the difference is between the two underlying proportions—provided it is nonzero—samples of sufficiently large size can virtually guarantee statistical significance. Assuming that an investigator desires to detect only differences that are of practical importance, and not merely differences of any magnitude, he should impose the added safeguard against a Type I error of not employing sample sizes that are larger than he needs to guard against the second kind of error.

The second kind of error, called the Type II error, consists in failing to declare the two proportions significantly different when in fact they are different. As just pointed out, such a failure is not necessarily an error when the proportions are only trivially different. It becomes an error only when the proportions differ to an important extent. The practical control over the Type II error must therefore begin with the investigator's specifying just what difference is of sufficient importance for him to desire to detect it, and must continue with his specifying the probability he desires of actually detecting it. This probability, denoted $1 - \beta$, is called the *power* of the test; the quantity β is the probability of failing to find the specified difference to be statistically significant.

Some means of specifying an important difference between proportions are given in Section 3.1. Having specified the quantities α, $1 - \beta$, and the minimum difference in proportions he deems important, the investigator may use the mathematical results of Section 3.2 or the values in Table A.3 (described in Section 3.3) to find the sample sizes necessary to assure that (1) any smaller sample sizes will reduce the chances below $1 - \beta$ of detecting the specified difference and (2) any larger sample sizes may increase the chances well above α of declaring a trivially small difference to be significant.

Frequently an investigator is restricted to working with sample sizes dictated by a prescribed budget or by a set time limit. He will still find the values in Table A.3 useful, for he can use them to find those differences that he has a reasonable probability of detecting and thus to obtain a realistic appraisal of the chances for success of his study.

Some additional points are made in Section 3.4.

3.1. SPECIFYING A DIFFERENCE WORTH DETECTING

An investigator will often have some idea of the order of magnitude of the proportions he is studying. This knowledge might come from his own or someone else's previous research, from an accumulation of clinical experience, from small-scale pilot work, or from vital statistics reports. Given at least some information, the investigator can, using his imagination and expertise, come up with an estimate of a difference between two proportions which is scientifically or clinically important. Given no information, the investigator has no basis for designing his study intelligently, and would be hard put to justify designing it at all.

In this section only two of the many approaches to the specification of a difference are illustrated, each with two examples. Let P_1 denote the proportion of members of the first group who possess the attribute or experience the outcome being studied. In general, the designation of one of the groups as the first and the other as the second is arbitrary. Here, however, we designate the first group to be the one that might be viewed as a standard, typically because more information may be available for it than for the other group. Our problem is to determine that value of P_2, the proportion in the second group, which, if actually found, would be deemed on practical grounds to differ sufficiently from P_1 to warrant the conclusion that the two groups are different.

Example 1. In a comparative clinical trial, the first group might represent patients treated by a standard form of therapy. The proportion P_1 might then refer to their observed response (e.g., remission) within a specified period of time following the beginning of treatment. The second group might represent patients treated with an as yet untested alternative form of therapy. A clinically important proportion P_2 associated with the alternative treatment might be determined as follows.

Suppose that it can be assumed that all patients responding to the standard treatment would also respond to the new therapy. Suppose further that, if at least an added fraction f, specified by the investigator, of nonresponders to the standard treatment respond to the new one, then he would wish to identify the new treatment as superior to the old. Since the proportion of non-responders to the standard treatment is $1 - P_1$, a clinically important value of P_2 is therefore $P_1 + f(1 - P_1)$.

For example, the remission rate associated with the standard treatment might be $P_1 = .60$. If the investigator will view the alternative treatment as superior to the standard only if it succeeds in remitting the symptoms of at least one quarter of those patients who would not otherwise show remission, so that $f = .25$, then he is in effect specifying a value $P_2 = .60 + .25 \times (1 - .60) = .70$ as one that is different to a practically important extent from $P_1 = .60$.

In this example, the proportions P_1 and P_2 refer to a favorable outcome, namely, a remission of symptoms. Similar reasoning can be applied to studies in which an untoward event (e.g., morbidity or mortality) is of especial interest.

Example 2. Suppose that the rate of premature births is P_1 among women of a certain age and race who attend the prenatal clinic in their community hospital. An intensive education program aimed at nonattenders is to be undertaken only if P_2, the rate of premature births among prospective mothers not attending the clinic but otherwise similar to the clinic attenders, is sufficiently greater than P_1.

It is reasonable to assume that a mother who delivered a premature offspring even after having attended the clinic would also have done so if she had not attended the clinic. The added risk of prematurity associated with nonattendance can thus only operate on mothers who do not deliver premature offspring after attending the clinic. If f denotes an added risk that is of practical importance, the hypothesized value of P_2 is then $P_1 + f(1 - P_1)$.

Suppose, for example, that the prematurity rate for clinic attenders is $P_1 = .25$. Suppose further that an education program is to be undertaken only if, of women who attend the clinic and who do not deliver a premature offspring, at least 20% would have delivered a premature offspring by not attending the clinic. The value of f is then .20, and the hypothesized value of P_2 is $.25 + .20 \times (1 - .25) = .40$.

The approach to the comparison of two proportions exemplified by these two examples has been recommended and applied by Sheps (1958, 1959, 1961). It is considered again in Chapter 7, where the *relative difference* $f = (P_2 - P_1)/(1 - P_1)$ is studied in greater detail.

Example 3. One often undertakes a study in order to replicate (or refute) another's research findings, or to see if one's own previous findings hold up in a new setting. One must be careful, however, to control for the possibility that the rates in the groups being compared are at levels in the new setting different from those in the old. This possibility effectively rules out attempting to recapture the simple difference between rates found previously.

For example, suppose that the rate of depression among women aged 20–49 was found in the mental hospitals of one community to be 40% higher than the rate among men aged 20–49. It will be impossible to find the same difference in the mental hospitals of a new community if, there, the rate of depression among males aged 20–49 is 70%, since a difference of 40% implies an impossible rate of 110% for women similarly aged.

A measure of the difference between two rates is therefore needed that may be expected to remain constant if the levels at which the rates apply vary across settings. A measure frequently found to have this property is the *odds ratio*, denoted ω. The odds ratio is discussed in greater detail in Chapters 5 and 6. Here we give only its definition.

If P_1 is the rate at which an event occurs in the first population, then the odds associated with that event in the first population are, say, $\Omega_1 = P_1/Q_1$, where $Q_1 = 1 - P_1$. Similarly, the odds associated with the event in the second population are $\Omega_2 = P_2/Q_2$. The odds ratio is simply the ratio of these two odds,

$$\omega = \frac{\Omega_2}{\Omega_1} = \frac{P_2 Q_1}{P_1 Q_2}. \tag{3.1}$$

The odds ratio is also termed the cross-product ratio (Fisher, 1962) and the approximate relative risk (Cornfield, 1951). If $P_2 = P_1$, then $\omega = 1$. If $P_2 < P_1$, then $\omega < 1$. If $P_2 > P_1$, then $\omega > 1$.

Suppose that a study is to be carried out in an attempt to replicate the results of a previous study in which the odds ratio was found to be ω. If, in the community in which the new study is to be conducted, the rate of occurrence of the event in the first group is P_1, and if the same value ω for the odds ratio is hypothesized to apply in the new community, then the value hypothesized for P_2 is

$$P_2 = \frac{\omega P_1}{\omega P_1 + Q_1}. \tag{3.2}$$

For example, suppose that the value $\omega = 2.5$ had previously been found as the ratio of the odds for depression among female mental hospital patients aged 20–49 to the odds for male mental hospital patients similarly aged. If the same value for the odds ratio is hypothesized to obtain in the mental hospitals of a new community, and if in that community's mental hospitals the rate of depression among male patients aged 20–49 is approximately $P_1 = .70$, then the rate among female patients aged 20–49 is hypothesized to be approximately

$$P_2 = \frac{2.5 \times .70}{2.5 \times .70 + .30} = .85.$$

An important property of the odds ratio to be demonstrated in Chapters 5 and 6 is that the same value should be obtained whether the study is a prospective or a retrospective one. This fact may be taken advantage of if an investigator wishes to replicate a previous study but alters the research design from, say, a retrospective to a prospective study.

Example 4. Suppose that a case-history (retrospective) study was conducted in a certain school district. School children with emotional disturbances requiring psychological care were compared with presumably normal children on a number of antecedent characteristics. Suppose it was found that one quarter of the emotionally disturbed children versus one tenth of the normal controls had lost (by death, divorce, or separation) at least one parent before age 5. The odds ratio is then, from (3.1),

$$\omega = \frac{.25 \times .90}{.10 \times .75} = 3.0.$$

Suppose that a study of this association is to be conducted prospectively in a new community by following through their school years a sample of children who begin school with both parents alive and at home (group 1) and a sample who begin school with at least one parent absent from the home (group 2), with the proportions developing emotional problems being compared.

From a survey of available school records, the investigator in the new school district is able to estimate that P_1, the proportion of children beginning school with both parents at home who ultimately develop emotional problems, is $P_1 = .05$. If the value $\omega = 3.0$ found in the retrospective study is hypothesized to apply in the new school district, the investigator is effectively hypothesizing a value (see equation 3.2)

$$P_2 = \frac{3.0 \times .05}{3.0 \times .05 + .95} = .136,$$

or approximately 15% as the rate of emotional disturbance during school years among children who have lost at least one parent before age 5.

The methods just illustrated may be of use in generating hypotheses for studies to be carried out within a relatively short time, but are likely to prove inadequate for long-term comparative studies. Halperin, Rogot, Gurian, and Ederer (1968) give a model and some numerical results when two long-term therapies are to be compared and when few or no dropouts are expected. When dropouts are likely to occur, the model of Schork and Remington (1967) may be useful for generating hypotheses.

3.2. THE MATHEMATICS OF SAMPLE SIZE DETERMINATION

We assume in this section and the next that the sample sizes from the two populations being compared, n_1 and n_2, are equal to a common n. We find the value for the common sample size n so that (1) if in fact there is no difference between the two underlying proportions, then the chance is α of falsely declaring the two proportions to differ; and (2) if in fact the proportions are P_1 and $P_2 \neq P_1$, then the chance is $1 - \beta$ of correctly declaring the two proportions to differ. Since this section only derives the mathematical results on which the values in Table A.3 (described in Section 3.3) are based, it may safely be left unread.

Suppose that the proportions found in the two samples are p_1 and p_2. The statistic used for testing the significance of their difference is, temporarily ignoring the continuity correction,

$$z = \frac{p_2 - p_1}{\sqrt{2\bar{p}\bar{q}/n}}, \tag{3.3}$$

where

$$\bar{p} = \tfrac{1}{2}(p_1 + p_2)$$

and

$$\bar{q} = 1 - \bar{p}.$$

To assure that the probability of a Type I error is α, the difference between p_1 and p_2 will be declared significant only if

$$|z| > c_{\alpha/2}, \tag{3.4}$$

where $c_{\alpha/2}$ denotes the value cutting off the proportion $\alpha/2$ in the upper tail of the standard normal curve and $|z|$ is the absolute value of z, always a nonnegative quantity. For example, if $\alpha = .05$, then $c_{.05/2} = c_{.025} = 1.96$, and the difference is declared significant if either $z > 1.96$ or $z < -1.96$.

If the difference between the underlying proportions is actually $P_2 - P_1$, we wish the chances to be $1 - \beta$ of rejecting the hypothesis, that is, of having the outcome represented in (3.4) actually occur. Thus we must find the value of n such that, when $P_2 - P_1$ is the difference between the proportions,

$$\Pr\left\{\frac{|p_2 - p_1|}{\sqrt{2\bar{p}\bar{q}/n}} > c_{\alpha/2}\right\} = 1 - \beta. \tag{3.5}$$

The probability in (3.5) is the sum of two probabilities,

$$1 - \beta = \Pr\left\{\frac{p_2 - p_1}{\sqrt{2\bar{p}\bar{q}/n}} > c_{\alpha/2}\right\} + \Pr\left\{\frac{p_2 - p_1}{\sqrt{2\bar{p}\bar{q}/n}} < -c_{\alpha/2}\right\}. \tag{3.6}$$

If P_2 is hypothesized to be greater than P_1, then the second probability on the right-hand side of (3.6)—representing the event that p_2 is appreciably less than p_1—is near zero (see Problem 3.1). Thus we need only find the value of n such that, when $P_2 - P_1$ is the actual difference,

$$1 - \beta = \Pr\left\{\frac{p_2 - p_1}{\sqrt{2\bar{p}\bar{q}/n}} > c_{\alpha/2}\right\}. \tag{3.7}$$

The probability in (3.7) cannot yet be evaluated because the mean and standard error of $p_2 - p_1$ appropriate when $P_2 - P_1$ is the actual difference have not yet been taken into account. The mean of $p_2 - p_1$ is $P_2 - P_1$, and its standard error is

$$\text{s.e.}(p_2 - p_1) = \sqrt{(P_1 Q_1 + P_2 Q_2)/n}, \tag{3.8}$$

where $Q_1 = 1 - P_1$ and $Q_2 = 1 - P_2$.

The following development of (3.7) can be traced using only simple algebra:

$$1 - \beta = \Pr\left\{(p_2 - p_1) > c_{\alpha/2}\sqrt{2\bar{p}\bar{q}/n}\right\}$$

$$= \Pr\left\{(p_2 - p_1) - (P_2 - P_1) > c_{\alpha/2}\sqrt{2\bar{p}\bar{q}/n} - (P_2 - P_1)\right\}$$

$$= \Pr\left\{\frac{(p_2 - p_1) - (P_2 - P_1)}{\sqrt{(P_1 Q_1 + P_2 Q_2)/n}} > \frac{c_{\alpha/2}\sqrt{2\bar{p}\bar{q}/n} - (P_2 - P_1)}{\sqrt{(P_1 Q_1 + P_2 Q_2)/n}}\right\}. \tag{3.9}$$

The final probability in (3.9) can be evaluated using tables of the normal distribution because, when the underlying proportions are P_2 and P_1, the quantity

$$Z = \frac{(p_2 - p_1) - (P_2 - P_1)}{\sqrt{(P_1 Q_1 + P_2 Q_2)/n}} \tag{3.10}$$

has, to a good approximation if n is large, the standard normal distribution.

Let $c_{1-\beta}$ denote the value cutting off the proportion $1 - \beta$ in the upper tail and β in the lower tail of the standard normal curve. Then, by definition,

$$1 - \beta = \Pr\{Z > c_{1-\beta}\}. \tag{3.11}$$

By matching (3.11) with the last probability of (3.9), we find that the value of n we seek is the one that satisfies

$$c_{1-\beta} = \frac{c_{\alpha/2}\sqrt{2\bar{p}\bar{q}/n} - (P_2 - P_1)}{\sqrt{(P_1 Q_1 + P_2 Q_2)/n}}$$

$$= \frac{c_{\alpha/2}\sqrt{2\bar{p}\bar{q}} - (P_2 - P_1)\sqrt{n}}{\sqrt{P_1 Q_1 + P_2 Q_2}}. \tag{3.12}$$

Before presenting the final expression for n, we note that (3.12) is a function not only of P_1 and P_2, which may be hypothesized by the investigator, but also of $\bar{p}\bar{q}$, which is observable only after the study is complete. If n is fairly large, however, $\bar{p}$ will be close to

$$\bar{P} = \frac{P_1 + P_2}{2}, \tag{3.13}$$

and, more importantly, $\bar{p}\bar{q}$ will be close to $\bar{P}\bar{Q}$, where $\bar{Q} = 1 - \bar{P}$. Therefore, replacing $\sqrt{2\bar{p}\bar{q}}$ in (3.12) by $\sqrt{2\bar{P}\bar{Q}}$ and solving for n, we find

$$n = \frac{(c_{\alpha/2}\sqrt{2\bar{P}\bar{Q}} - c_{1-\beta}\sqrt{P_1 Q_1 + P_2 Q_2})^2}{(P_2 - P_1)^2} \tag{3.14}$$

to be the required sample size from *each* of the two populations being compared when the continuity correction is not employed.

Kramer and Greenhouse (1959) derived a formula for sample sizes when the continuity correction is incorporated into the test statistic. Let n be defined as in (3.14), and let n' denote the sample size required in each group when the continuity correction is employed. Then

$$n' = \frac{n}{4}[1 + \sqrt{1 + 8/(n|P_2 - P_1|)}]^2. \tag{3.15}$$

3.3. USING THE SAMPLE SIZE TABLES

Table A.3 gives the sample sizes necessary in each of the two groups being compared for varying values of the hypothesized proportions P_1 and P_2, for varying significance levels ($\alpha = .01, .02, .05, .10,$ and $.20$), and for varying powers ($1 - \beta = .50, .65(.05).95, .99$). The value $1 - \beta = .50$ is included not so much because an investigator will intentionally embark on a study for which the chances of success are only $50:50$, but rather to help provide him with a baseline for the minimum sample sizes necessary.

The probability of a Type I error, α, is frequently specified first. If, on the basis of declaring the two proportions to differ significantly, the decision is made to conduct further (possibly expensive) research or to replace a standard form of treatment with a new one, then the Type I error is serious and α should be kept small (say, .01 or .02). If the study is aimed only at adding to the body of published knowledge concerning some theory, then the Type I error is less serious, and α may be increased to .05 or .10 (the more the published evidence points to a difference, the higher may α safely be set).

Having specified α, the investigator needs next to specify the chances $1 - \beta$ of detecting the proportions as different if, in the underlying populations, the proportions are P_1 and P_2. The criterion suggested by Cohen (1969, p. 54) seems reasonable. He supposes it to be the typical case that a Type I error is some four times as serious as a Type II error. This implies that one should set β, the probability of a Type II error, approximately equal to 4α, so that the power becomes, approximately, $1 - \beta = 1 - 4\alpha$. Thus when $\alpha = .01$, $1 - \beta$ may be set at .95; for $\alpha = .02$, set $1 - \beta = .90$; and for $\alpha = .05$, set $1 - \beta = .80$. When α is larger than .05, it seems safe to take $1 - \beta = .75$ or less. The use of Table A.3 will be illustrated for each of the examples of Section 3.1.

Example 1.   The investigator hypothesizes a remission rate of $P_1 = .60$ for the standard treatment and one of $P_2 = .70$ for the new treatment. He sets his significance level α at .01 and his power $1 - \beta$ at .95. He finds that he needs to study 847 patients under the standard treatment and 847 under the new one, the assignment of patients to treatment groups being at random, in order to guarantee his desired significance level and power.

If he is willing to reduce his chances of detecting a difference to $1 - \beta = .75$, but not to increase his significance level, he finds that he would need to study 519 patients with each treatment. If he can afford to study no more than a total of 600 patients, so that each treatment would be applied to no more than 300 subjects, his chances of detecting the hypothesized difference become less than $50:50$.

Example 2.   The investigator hypothesizes a prematurity rate of $P_1 = .25$ for clinic attenders and one of $P_2 = .40$ for nonattenders. He sets his significance level α at .01 and his power $1 - \beta$ at .95. He finds that he needs to study 370 mothers from each group, all women being within a specified age range. If he increases his significance level to $\alpha = .02$ and lowers his power to $1 - \beta = .90$, he finds that he needs 278 mothers from each group.

Example 3. The investigator hypothesizes the rate of depression among male mental hospital patients aged 20–49 to be $P_1 = .70$, and the rate among similarly aged female patients to be $P_2 = .85$. He sets his significance level α at .05 and his power $1 - \beta$ at .80. He finds that he needs to study 146 patients of each sex. If the investigator had planned to study 250 patients of each sex, his chances of picking up the hypothesized difference would be over 95%, a value that might be larger than he needs.

Example 4. The investigator hypothesizes that the proportion developing emotional problems among children beginning school with both parents at home is $P_1 = .05$, and that the proportion among children beginning school with at least one parent absent is approximately $P_2 = .15$. He sets his significance level α at .05 and his power $1 - \beta$ at .80. He finds that he needs to follow up 178 of each kind of child, making certain that the two cohorts are similar with respect to sex and race. If he can afford to study no more than 120 of each kind of child, and if he is willing to increase his chance of making a Type I error to $\alpha = .10$, he will still have about a 70% chance of finding the groups to be different if his hypothesized values of P_1 and P_2 are correct.

3.4. SOME ADDITIONAL COMMENTS

Cohen (1969, Chapter 6) gives a set of tables for determining sample sizes when the same parameters as above are specified. Since the significance test he considers is different from the standard one, and does not incorporate the continuity correction, the sample sizes in his tables are slightly different from those in Table A.3 here. In general, Cohen's tables may be used if the investigator can hypothesize the order of magnitude of the difference between P_1 and P_2, but not their separate magnitudes. If the investigator can hypothesize the separate values of P_1 and P_2, the current table is preferable.

It has so far been assumed that a two-tailed test (see Section 2.3) would be used in comparing the two proportions. If the investigator chooses to perform a one-tailed test, he can still use Table A.3, but should enter it with twice his significance level. Thus if he desires a one-tailed significance level of .01, he should enter the table with $\alpha = .02$; if he desires a one-tailed significance level of .05, he should enter the table with $\alpha = .10$. He need make no change in the value of $1 - \beta$.

Problem 3.1. Suppose that $P_2 > P_1$ and that n is the sample size studied in each group. Let Z represent a random variable having the standard normal distribution.

(a) Show that the probability that p_2 is significantly *less* than p_1,

$$\Pr\left\{\frac{p_2 - p_1}{\sqrt{2\bar{p}\bar{q}/n}} < -c_{\alpha/2}\right\},$$

is approximately equal to

$$\Pi = \Pr\left\{Z < \frac{-c_{\alpha/2}\sqrt{2\bar{P}\bar{Q}} - (P_2 - P_1)\sqrt{n}}{\sqrt{P_1Q_1 + P_2Q_2}}\right\}.$$

(b) If $P_2 = P_1$, then $\Pi = \alpha/2$. Thus, if $P_2 > P_1$, show why $\Pi < \alpha/2$. (*Hint.* Prove that $\sqrt{P_1Q_1 + P_2Q_2} < \sqrt{2\bar{P}\bar{Q}}$ whenever $P_2 \neq P_1$. Therefore, if $P_2 > P_1$,

$$\frac{-c_{\alpha/2}\sqrt{2\bar{P}\bar{Q}} - (P_2 - P_1)\sqrt{n}}{\sqrt{P_1Q_1 + P_2Q_2}} < -c_{\alpha/2} - \frac{(P_2 - P_1)\sqrt{n}}{\sqrt{P_1Q_1 + P_2Q_2}} < -c_{\alpha/2}.)$$

(c) Π is small even if P_2 is only slightly larger than P_1 and even if n is small. Find the value of Π when $P_1 = .10$, $P_2 = .11$, $n = 9$, and $\alpha = .05$. Note that the probability found in Table A.2 must be *halved*.

Problem 3.2. Let the notation and assumptions of Problem 3.1 be used again. The power of the test for comparing p_1 and p_2 is approximately

$$1 - \beta = \Pr\left\{Z > \frac{c_{\alpha/2}\sqrt{2\bar{P}\bar{Q}} - (P_2 - P_1)\sqrt{n}}{\sqrt{P_1Q_1 + P_2Q_2}}\right\}.$$

(a) Show that $1 - \beta$ approaches unity as n becomes large but α remains fixed. (*Hint.* What is the probability that a standard normal random variable exceeds -1? -2? -3? What value does the expression to the right of the inequality sign above approach as n increases?)
(b) Show that $1 - \beta$ decreases as α becomes small but n remains fixed.

Problem 3.3. An investigator hypothesizes that the improvement rate associated with a placebo is $P_1 = .45$, and that the improvement rate associated with an active drug is $P_2 = .65$. He plans to perform a *one-tailed* test.
(a) If he desires a significance level of $\alpha = .01$ and a power of $1 - \beta = .95$, how large a sample per treatment must he study?
(b) How large must his sample sizes be if he relaxes his significance level to $\alpha = .05$ and his power to $1 - \beta = .80$?

REFERENCES

Cohen, J. (1969). *Statistical power analysis for the behavioral sciences.* New York: Academic Press.

Cornfield, J. (1951). A method of estimating comparative rates from clinical data. Applications to cancer of the lung, breast and cervix. *J. natl. Cancer Inst.,* **11,** 1269–1275.

Fisher, R. A. (1962). Confidence limits for a cross-product ratio. *Austral. J. Statist.,* **4,** 41.

Halperin, M., Rogot, E., Gurian, J. and Ederer, F. (1968). Sample sizes for medical trials with special reference to long-term therapy. *J. chronic Dis.,* **21,** 13–24.

Kramer, M. and Greenhouse, S. W. (1959). Determination of sample size and selection of cases. Pp. 356–371 in National Academy of Sciences—National Research Council Publication 583, *Psychopharmacology: Problems in evaluation.* Washington, D.C.

Schork, M. A. and Remington, R. D. (1967). The determination of sample size in treatment-control comparisons for chronic disease studies in which drop-out or non-adherence is a problem. *J. chronic Dis.,* **20,** 233–239.

Sheps, M. C. (1958). Shall we count the living or the dead? *New Engl. J. Med.,* **259,** 1210–1214.

Sheps, M. C. (1959). An examination of some methods of comparing several rates or proportions. *Biometrics,* **15,** 87–97.

Sheps, M. C. (1961). Marriage and mortality. *Amer. J. public Health,* **51,** 547–555.

CHAPTER 4

How to Randomize

A number of references have been made so far to randomization. Subjects were described as being "randomly selected" from a larger group of subjects, as being "randomly assigned" to one or another treatment group, and so on. This chapter gives some methods for achieving randomness of selection (needed in sampling methods I and II; see Section 4.1) or of assignment (needed in sampling method III; see Section 4.2).

It is important to bear in mind that randomness inheres not in the samples one ends up with, but in the method used to generate those samples. When we say that a group is a *simple random sample* from a larger group, we mean that each possible sample has the same chance of being selected. When we say that treatments are assigned to subjects *at random*, we mean that each subject is equally likely to receive each of the treatments.

The necessity for randomization in controlled experiments was first pointed out by Fisher (1935). In the context of comparative trials, Hill (1962) describes what is accomplished by the random assignment of treatments to subjects:

> [Randomization] ensures three things: it ensures that neither our personal idiosyncracies . . . nor our lack of balanced judgement has entered into the construction of the different treatment groups . . . ; it removes the danger, inherent in an allocation based upon personal judgement, that believing we may be biased in our judgements we endeavour to allow for that bias, to exclude it, and that in so doing we may overcompensate and by thus "leaning over backward" introduce a lack of balance from the other direction; and, having used a random allocation, the sternest critic is unable to say when we eventually dash into print that quite probably the groups were differentially biased through our predilections or through our stupidity [p. 35].

Table A.4 presents 20,000 random digits, arrayed on each page in ten columns of 50 numbers, with five digits to each number. Some illustrations of the use of the table follow.

4.1. SELECTING A SIMPLE RANDOM SAMPLE

Suppose that a firm has 250 employees, of whom 100 are to be selected for a thorough physical examination and an interview to determine health habits. A simple random sample of 100 out of the larger group of 250 may be selected as follows.

Examine consecutive three digit numbers, ignoring any that are 000 or between 251 and 999. Of the numbers between 001 and 250, list the first 100 distinct ones that are encountered. When a column is completed, proceed to the next one. The 100 numbers listed designate the employees to be selected. If a card file exists containing the names of all employees—the order in which the names appear is immaterial—the employees may be numbered from 1 to 250 and those whose numbers appear on the list of random numbers would be selected.

To illustrate, let us begin in the second column of the second page of Table A.4. Each number in the column contains five digits, of which only the first three will be examined. The first five numbers in the column, after deleting their last two digits, are 670, 716, 367, 988, and 283—all greater than 250. The sixth number, 142, is between 001 and 250 and thus designates one of the employees to be selected. The other numbers selected from the second column are seen to be 021, 166, 127, 060, 098, 219, 161, 042, 043, 157, 113, 234, 024, 028, and 128.

Having exhausted the second column, and still requiring 84 additional numbers, we proceed to examine the third column of the second page. At the end of the third column, 29 distinct numbers between 001 and 250 are available. In numerical order, they are

001	028	052	107	142	166
014	034	059	113	146	219
021	042	060	121	157	234
024	043	080	127	160	244
026	047	098	128	161	

Subsequent columns are examined similarly until an additional 71 distinct numbers between 001 and 250 are found. If a previously selected number is encountered (e.g., 244 is encountered again in column 4), it is ignored.

4.2. RANDOMIZATION IN A CLINICAL TRIAL

Suppose that a clinical trial is to be carried out to compare the effectiveness of a drug with that of an inert placebo, and suppose that 50 patients are to be studied under each drug, requiring a total of 100 patients. Suppose, finally, that patients enter the study serially over time, and so are not all available at once.

Two randomization methods exist. The first calls for selecting 50 distinct numbers between 001 and 100, as described in Section 4.1, and for letting these numbers designate the patients who will receive the active drug. The remaining 50 numbers designate the patients who will receive the placebo.

This method has two drawbacks. First, if the study must be terminated prematurely, there exists a strong likelihood that the total number of patients who had been assigned the active drug up to the termination date will not equal the total number who had been assigned the placebo. Statistical comparisons lose sensitivity if the sample sizes differ. Second, if the clinical characteristics of the patients entering the trial during one interval of time differ from those of patients entering during another, or if the standards of assessment change over time, then the two treatment groups might well end up being different, in spite of randomization, either in the kinds of patients they contain or in the standards of assessment applied to them (see Cutler, Greenhouse, Cornfield, and Schneiderman, 1966, p. 865).

The second possible method of randomization guards against these potential weaknesses in the first method. It calls for independently randomizing, to one treatment group or the other, patients who enter the trial within each successive short interval of time.

Suppose, for example, that ten patients are expected to enter the trial each month. A reasonable strategy is to assign at random five of the first ten patients to one treatment group and the other five to the second treatment group, and to repeat the random assignment of five patients to one group and five to the other for each successive group of ten.

The procedure is implemented as follows, beginning at the top of the fifth column of the first page of Table A.4. Because selection is from ten subjects, only single digits need be examined, with 0 designating the tenth subject. The first five distinct digits are found to be 2, 5, 4, 8, and 6. Therefore, the second, fourth, fifth, sixth, and eighth patients out of the first ten will be assigned the active drug, and the others—the first, third, seventh, ninth, and tenth—will be assigned the placebo.

Examination of the column continues for the second series of ten patients. The next five distinct digits are found to be 3, 1, 8, 0, and 5, implying that, of the second group of ten patients, the first, third, fifth, eighth, and tenth are assigned the active drug and the second, fourth, sixth, seventh, and ninth are assigned the placebo. As soon as the leading digits in a column are exhausted, the second digits may be examined.

It is important that a new set of random numbers be selected for each successive group, lest an unsuspected periodicity in the kind of patient entering the trial, or a pattern soon apparent to personnel who should be kept ignorant of which drugs the patients are receiving, introduce a bias.

A special case of the method just illustrated is the pairing of subjects, with one member of the pair randomly assigned the active drug and the other assigned the placebo. Random assignment becomes especially simple when subjects are paired. To begin with, one member of the pair must be designated the first and the other, the second. The designation might be on the basis of time of entry into the study, of alphabetical order of the surname, or of any other criterion, provided that the designation is made before the randomization is performed.

Table A.4 is entered at any convenient point, and successive single digits are examined, one digit for each pair. If the digit is odd—1, 3, 5, 7, or 9—the first member of the pair is assigned the active drug and the second member, the placebo. If the digit is even—2, 4, 6, 8, or 0—the second member of the pair is assigned the active drug and the first member the placebo.

To illustrate, let us begin at the top of the first column on the third page of Table A.4. The first digit encountered, 2, is even, indicating that, in the first pair, the active drug is given to the second member and the placebo to the first. The second digit encountered, 8, is also even, so that, in the second pair, too, the active drug is given to the second member and the placebo to the first. The sixth, seventh, and eighth digits—3, 9, and 1—are all odd. Therefore, in the sixth through eighth pairs, the active drug is given to the first member and the placebo to the second.

Investigators who require more than the 20,000 random digits of Table A.4 are referred to the Rand Corporation's extensive table (1955). Those whose research designs call for applying each of a number (more than two) of treatments to each of a sample of subjects will find Moses and Oakford's tables of random permutations (1963) indispensable. Their tables also facilitate each of the uses of randomization illustrated above.

REFERENCES

Cutler, S. J., Greenhouse, S. W., Cornfield, J. and Schneiderman, M. A. (1966). The role of hypothesis testing in clinical trials: Biometrics seminar. *J. chronic Dis.*, **19**, 857–882.

Fisher, R. A. (1935). *The design of experiments*. Edinburgh: Oliver and Boyd.

Hill, A. B. (1962). *Statistical methods in clinical and preventive medicine*. New York: Oxford University Press.

Moses, L. E. and Oakford, R. V. (1963). *Tables of random permutations*. Stanford: Stanford University Press.

Rand Corporation (1955). *A million random digits with* 100,000 *normal deviates*. New York: The Free Press.

CHAPTER 5

Sampling Method I: Naturalistic or Cross-Sectional Studies

In this chapter we study what had been identified in Section 2.1 as method I sampling. This method of sampling, referred to as *cross-sectional*, *naturalistic*, or *multinomial* sampling, does not attempt to prespecify any frequencies except the overall total.

We consider only the case where the resulting data are arrayed in a 2 × 2 table. Most statistics texts describe the chi square test for association when there are more than two rows or more than two columns in the resulting cross-classification table (e.g., Dixon and Massey, 1957, Section 13.3; and Maxwell, 1961, Chapter 2). The accuracy of the chi square test for general contingency tables when the total sample size is small has been studied by Craddock and Flood (1970). Methods for estimating association in such tables are given by Goodman and Kruskal (1954, 1959), Goodman (1964), and Altham (1970a, 1970b).

In Section 5.1 we present some hypothetical data that are referred to repeatedly in this and the next chapter. In Section 5.2 we examine the estimation by means of measures based on χ^2 of the degree of association between the two characteristics studied, and examine other measures in Section 5.3. Some properties of the odds ratio and of its logarithm are presented in Section 5.4.

5.1. SOME HYPOTHETICAL DATA

Suppose that we are studying the association, if any, between the age of the mother (A represents a maternal age less than or equal to 20 years; $\bar{A}$, a maternal age over 20 years) and the birthweight of her offspring (B represents a birthweight less than or equal to 2500 grams; $\bar{B}$, a birthweight over 2500 grams). Since the association might vary as a function of race and socio-economic status, let us agree to study only black women from social classes IV and V who deliver in a single specified hospital.

Suppose that all the data we need are on file in the hospital's record room and that, for each delivery, a card is punched recording the birthweight of the infant and the age, color, and social class of the mother. After sorting out the cards for the mothers who do not fit the study criteria, we are left with a file of cards for deliveries to black mothers of social classes IV and V. Suppose that the number of cards remaining is quite large and that the determination of A versus $\bar{A}$ or of B versus $\bar{B}$ is not too simple. For example, the mother's date of birth might be recorded instead of her age, making card sorting complicated. The decision might then be made to examine only a sample, say 200, of all the records.

The sample should ideally be a simple random sample (see Chapter 4), but alternatives exist. Suppose that there are a total of 1000 records. A *systematic random sample* of 200 may be selected by drawing every fifth record, with the starting record (the first, second, third, fourth, or fifth) chosen at random. Another alternative is to base the selection on the last digit of the identification number, choosing only those records whose last digit is, for example, a 3 or a 7.

Table 5.1. *Association between birthweight and maternal age: cross-sectional study*

	Birthweight		
Maternal Age	B	$\bar{B}$	Total
A	10	40	50
$\bar{A}$	15	135	150
Total	25	175	200

Let us suppose that the sample of 200 records has been selected, and that the data are as in Table 5.1. As pointed out in Chapter 2, a more appropriate means of presenting the data resulting from method I sampling is as in Table 5.2.

Table 5.2. *Joint proportions derived from Table 5.1*

	Birthweight		
Maternal Age	B	$\bar{B}$	Total
A	$.05 (= p_{11})$	$.20 (= p_{12})$	$.25 (= p_{1.})$
$\bar{A}$	$.075 (= p_{21})$	$.675 (= p_{22})$	$.75 (= p_{2.})$
Total	$.125 (= p_{.1})$	$.875 (= p_{.2})$	$1.$

A consequence of sampling method I is that all probabilities may be estimated. Thus the proportion of all deliveries in which the mother was aged 20 years or less and in which the offspring weighed 2500 grams or less is estimated as $p(A \text{ and } B) = p_{11} = .05$. The proportion of all deliveries in which the mother was aged 20 years or less is estimated as $p(A) = p_{1.} = .25$, and the proportion of all deliveries in which the offspring weighed 2500 grams or less is $p(B) = p_{.1} = .125$.

The significance of the association between maternal age and birthweight (the first, but by no means the most important, issue) may be assessed by means of the standard chi square test. The value of the test statistic is

$$\chi^2 = \frac{200(|10 \times 135 - 40 \times 15| - \frac{1}{2}200)^2}{50 \times 150 \times 25 \times 175} = 2.58, \tag{5.1}$$

indicating an association that is not statistically significant.

5.2. MEASURES OF ASSOCIATION BASED ON χ^2

The failure to find statistical significance would presumably signal the completion of the analysis. For later comparative purposes, however, we proceed to consider the estimation of the *degree* of association between the two characteristics, beginning with estimates based on the magnitude of χ^2.

A common mistake is to use the value of χ^2 itself as the measure of association. Even though χ^2 is excellent as a measure of the *significance* of the association, it is not at all useful as a measure of the *degree* of association. The reason is that χ^2 is a function both of the proportions in the various cells and of the total number of subjects studied. The degree of association present is really only a function of the cell proportions. The number of subjects studied plays a role in the chances of finding significance if association exists, but should play no role in determining the extent of association (see, e.g., Fisher, 1954, pp. 89–90).

Suppose, for example, that another investigator studied the characteristics of 400 births from the same hospital, and suppose that the resulting data were as in Table 5.3. The value of χ^2 for these data is

$$\chi^2 = \frac{400(|20 \times 270 - 80 \times 30| - \frac{1}{2}400)^2}{100 \times 300 \times 50 \times 350} = 5.97,$$

which indicates an association significant at the .05 level.

The inferences are different for the data of Tables 5.1 and 5.3: nonsignificance for the first but significance for the second. The only reason for the difference, however, is that twice as many births were used in Table 5.3 as in Table 5.1. The joint proportions (see Table 5.2) are obviously identical

Table 5.3. Association between birthweight and maternal age: sampling method I with 400 births

	Birthweight		
Maternal Age	B	$\bar{B}$	Total
A	20	80	100
$\bar{A}$	30	270	300
Total	50	350	400

for Tables 5.1 and 5.3, so that the associations between maternal age and birthweight implied by both are also identical. The larger value of χ^2 for the data of Table 5.3 is thus a reflection only of the larger total sample size (the doubling of all frequencies has in fact more than doubled the value of chi square), and not of a greater degree of association.

A measure of the degree of association between characteristics A and B which is derived from χ^2 but is free of the influence of the total sample size, $n_{..}$, is the *phi coefficient:*

$$\varphi = \sqrt{\frac{\chi_u^2}{n_{..}}}, \tag{5.2}$$

where χ_u^2 is the uncorrected chi square statistic,

$$\chi_u^2 = \frac{n_{..}(n_{11}n_{22} - n_{12}n_{21})^2}{n_{1.}n_{2.}n_{.1}n_{.2}}. \tag{5.3}$$

The phi coefficient is especially popular as a measure of association in the behavioral sciences, and is interpretable as a correlation coefficient (Nunnally, 1967, pp. 118–120). Values close to zero indicate little if any association, whereas values close to unity indicate almost perfect predictability: if φ is near 1, then knowing whether a subject is A or $\bar{A}$ permits an accurate prediction of whether the subject is B or $\bar{B}$. The maximum value of φ is unity (if the marginal distributions are not equal, the maximum is actually less than unity) and, as a rule of thumb, any value less than .30 or .35 may be taken to indicate no more than trivial association.

For the data of Table 5.1,

$$\chi_u^2 = \frac{200(10 \times 135 - 40 \times 15)^2}{50 \times 150 \times 25 \times 175} = 3.43,$$

so that

$$\varphi = \sqrt{\frac{3.43}{200}} = .13, \tag{5.4}$$

hardly of appreciable magnitude. The value of φ for the data of Table 5.3 is obviously also equal to .13.

The phi coefficient finds its greatest usefulness in the study of items contributing to psychological and educational tests (Lord and Novick, 1968, Section 15.5) and in the *factor analysis* of a number of yes-no items. See Harman (1960) and Nunnally (1967, Chapters 9–10) for a general description of factor analysis, and Nunnally (1967, p. 308) and Lord and Novick (1968, p. 349) for the validity of factor analysis when applied to phi coefficients. Berger (1961) has presented a method for comparing phi coefficients from two independent studies.

The phi coefficient has a number of serious deficiencies, however. As shown in Chapter 6, the values of φ obtained when the association between characteristics A and B is studied prospectively and retrospectively are not comparable, nor is either value comparable to that obtained when the association is studied naturalistically. Carroll (1961) has shown that if either or both characteristics are dichotomized by cutting a continuous distribution into two parts, then the value of φ depends strongly on where the cutting point is set.

This lack of invariance of the phi coefficient and of other measures derived from χ^2 (see Goodman and Kruskal, 1954, pp. 739–740), plus presumably other reasons, led Goodman and Kruskal to assert that they "have been unable to find any convincing published defense of χ^2–like statistics as measures of association (1954, p. 740)." Whereas this assertion ignores the usefulness of φ in psychometrics, it does point to the avoidance of φ and of other statistics based on χ^2 as measures of association in those areas of research where comparability of findings but not necessarily of methods is essential.

5.3. OTHER MEASURES OF ASSOCIATION: THE ODDS RATIO

Goodman and Kruskal (1954, 1959) present a great many measures of association for 2×2 tables that are not functions of χ^2, and give their statistical properties in a later paper (1963). Here we concentrate on one such measure, the *odds ratio*.

Frequently, one of the two characteristics being studied is antecedent to the other. In the example we have been considering, maternal age is antecedent to birthweight. A measure of the risk of experiencing the outcome under study when the antecedent factor is present is

$$\Omega_A = \frac{P(B \mid A)}{P(\bar{B} \mid A)} \tag{5.5}$$

(see Section 1.1 for a definition of conditional probabilities). Ω_A is the *odds* that B will occur when A is present. Since $P(B \mid A)$ may be estimated by

$$p(B \mid A) = \frac{p_{11}}{p_{1.}}$$

and $P(\bar{B} \mid A)$ by

$$p(\bar{B} \mid A) = \frac{p_{12}}{p_{1.}},$$

therefore Ω_A may be estimated by

$$O_A = \frac{p_{11}/p_{1.}}{p_{12}/p_{1.}} = \frac{p_{11}}{p_{12}}. \tag{5.6}$$

For our example, the estimated odds that a mother aged 20 years or less will deliver an offspring weighing 2500 grams or less are, from Table 5.2,

$$O_A = \frac{.05}{.20} = \frac{1}{4} = .25. \tag{5.7}$$

Thus, for every four births weighing over 2500 grams to mothers aged 20 years or less, there is one birth weighing 2500 grams or less.

The information conveyed by these odds is exactly the same as that conveyed by the rate of low birthweight specific to young mothers,

$$p(B \mid A) = \frac{.05}{.25} = \frac{1}{5} = .20,$$

but the emphases differ. One can imagine attempting to educate prospective mothers aged 20 years or less. The impact of the statement, "One out of every five of you is expected to deliver an infant with a low birthweight," may well be different from the impact of "For every four of you who deliver infants of fairly high weight, one is expected to deliver an infant of low birthweight."

When A is absent, the odds of B's occurrence are defined as

$$\Omega_{\bar{A}} = \frac{P(B \mid \bar{A})}{P(\bar{B} \mid \bar{A})}, \tag{5.8}$$

which may be estimated as

$$O_{\bar{A}} = \frac{p_{21}/p_{2.}}{p_{22}/p_{2.}} = \frac{p_{21}}{p_{22}}. \tag{5.9}$$

For our example, the estimated odds that a mother aged more than 20 years will deliver an offspring weighing 2500 grams or less are

$$O_{\bar{A}} = \frac{.075}{.675} = \frac{1}{9} = .11. \tag{5.10}$$

Thus for every nine births weighing over 2500 grams to mothers aged more than 20 years (as opposed to every four to younger mothers), there is one birth weighing 2500 grams or less.

The two odds, Ω_A (5.5) and $\Omega_{\bar{A}}$ (5.8), may be contrasted in a number of ways in order to provide a measure of association. One such measure, due to Yule (1900), is

$$Q = \frac{\Omega_A - \Omega_{\bar{A}}}{\Omega_A + \Omega_{\bar{A}}}. \tag{5.11}$$

Another, also due to Yule (1912), is

$$Y = \frac{\sqrt{\Omega_A} - \sqrt{\Omega_{\bar{A}}}}{\sqrt{\Omega_A} + \sqrt{\Omega_{\bar{A}}}}. \tag{5.12}$$

The measure of association based on Ω_A and $\Omega_{\bar{A}}$ that is currently in greatest use is simply their ratio,

$$\omega = \frac{\Omega_A}{\Omega_{\bar{A}}}, \tag{5.13}$$

which may be estimated by the sample odds ratio,

$$o = \frac{O_A}{O_{\bar{A}}} = \frac{p_{11}/p_{12}}{p_{21}/p_{22}} = \frac{p_{11}p_{22}}{p_{12}p_{21}}. \tag{5.14}$$

If the two rates $P(B \mid A)$ and $P(B \mid \bar{A})$ are equal, indicating the independence or lack of association between the two characteristics, then the two odds Ω_A and $\Omega_{\bar{A}}$ are also equal (see Problem 5.1) so that the odds ratio $\omega = 1$. If $P(B \mid A) > P(B \mid \bar{A})$, then $\Omega_A > \Omega_{\bar{A}}$ and $\omega > 1$ (see Problem 5.2). If $P(B \mid A) < P(B \mid \bar{A})$, then $\Omega_A < \Omega_{\bar{A}}$ and $\omega < 1$.

For our data, the estimated odds ratio is

$$o = \frac{.05 \times .675}{.20 \times .075} = 2.25, \tag{5.15}$$

indicating that the odds of a young mother's delivering an offspring with low birthweight are $2\frac{1}{4}$ times those for an older mother. Because the odds ratio may also be estimated as

$$o = \frac{n_{11}n_{22}}{n_{12}n_{21}}, \tag{5.16}$$

it is sometimes also referred to as the *cross-product ratio*.

The standard error of the estimated odds ratio is approximately

$$\text{s.e.}(o) = \frac{o}{\sqrt{n_{..}}} \sqrt{\frac{1}{p_{11}} + \frac{1}{p_{12}} + \frac{1}{p_{21}} + \frac{1}{p_{22}}}. \tag{5.17}$$

For the data of Table 5.2, the standard error is found to be

$$\text{s.e.}(o) = \frac{2.25}{\sqrt{200}}\sqrt{\frac{1}{.05} + \frac{1}{.20} + \frac{1}{.075} + \frac{1}{.675}} = 1.00. \tag{5.18}$$

An equivalent formula in terms of the original frequencies is

$$\text{s.e.}(o) = o\sqrt{\frac{1}{n_{11}} + \frac{1}{n_{12}} + \frac{1}{n_{21}} + \frac{1}{n_{22}}}. \tag{5.19}$$

The standard error is useful in gauging the precision of the estimated odds ratio, but not in testing its significance. The classic chi square test should be used as a test of the hypothesis that the odds ratio in the population is equal to 1.

Anscombe (1956), Gart (1966), and Gart and Zweifel (1967) have studied the sampling properties of o and its standard error. Note that if either n_{12} or n_{21} is equal to zero, then o (5.16) is undefined. If any one of the four cell frequencies is equal to zero, then s.e.(o) (5.19) is undefined. Suggested improved estimates are

$$o' = \frac{(n_{11} + .5)(n_{22} + .5)}{(n_{12} + .5)(n_{21} + .5)} \tag{5.20}$$

for the odds ratio and

$$\text{s.e.}(o') = o'\sqrt{\frac{1}{n_{11} + .5} + \frac{1}{n_{12} + .5} + \frac{1}{n_{21} + .5} + \frac{1}{n_{22} + .5}} \tag{5.21}$$

for its standard error.

A number of important properties of the odds ratio as a measure of association will be demonstrated in the sequel. Advantages of using the odds ratio instead of other measures have been illustrated by Mosteller (1968). Edwards (1963) considered the advantages to be so great that he recommended that only the odds ratio or functions of it be used to measure association in 2 × 2 tables.

5.4. SOME PROPERTIES OF THE ODDS RATIO AND ITS LOGARITHM

The odds ratio was originally proposed by Cornfield (1951) as a measure of the degree of association between an antecedent factor and an outcome event such as morbidity or mortality, but only because it provided a good approximation to another measure he proposed, the *relative risk*. If the risk of the occurrence of event B when A is present is taken simply as the rate of

B's occurrence specific to the presence of A, $P(B \mid A)$, and similarly for the risk of B when A is absent, then the relative risk is simply the ratio of the two risks,

$$R = \frac{P(B \mid A)}{P(B \mid \bar{A})}. \tag{5.22}$$

R may be estimated by

$$r = \frac{p_{11}/p_{1.}}{p_{21}/p_{2.}} = \frac{p_{11}p_{2.}}{p_{21}p_{1.}}. \tag{5.23}$$

If the occurrence of event B is unlikely, whether or not characteristic A is present, then, as shown in Problem 5.3, r is approximately equal to o (5.14). For the data of Table 5.2,

$$r = \frac{.05 \times .75}{.075 \times .25} = 2.0, \tag{5.24}$$

only slightly less than the value found for the odds ratio, $o = 2.25$ (5.15).

There is more to the odds ratio, however, than merely an approximation to the relative risk. There exists a mathematical model, the so-called *logistic model*, which naturally gives rise to the odds ratio as a measure of association. Consider, for specificity, the association between cigarette smoking and lung cancer. Mortality from lung cancer is a function not only of whether one smokes but also, as but one example, of the amount of air pollution in the environment of the community where he works or lives.

Let us agree to study the association between smoking and lung cancer in one community only, and let x represent the mean amount of a specified pollutant in the atmosphere surrounding that community. A possible representation of the mortality rate from lung cancer for cigarette smokers is

$$P_S = \frac{1}{1 + e^{-(ax+b_S)}}, \tag{5.25}$$

and of the mortality rate for nonsmokers,

$$P_N = \frac{1}{1 + e^{-(ax+b_N)}}, \tag{5.26}$$

where $e = 2.718$, the base of natural logarithms. The parameter a measures the dependence of mortality on the specified air pollutant. The use of the same parameter, a, in (5.25) and (5.26) is equivalent to the assumption of no synergistic effect of smoking and air pollution on mortality. If a is positive, then both P_S and P_N approach unity as x, the mean amount of the pollutant, becomes large.

According to the model represented by (5.25) and (5.26), the effect of smoking on mortality is reflected only in the possible difference between the parameters b_S and b_N. When $x = 0$, that is, when a community is completely free of the specified pollutant, then b_S is directly related to the mortality rate for smokers and b_N is directly related to the mortality rate for nonsmokers.

Consider, now, the odds that a smoker from the selected community will die of lung cancer. These odds are

$$\Omega_S = \frac{P_S}{1 - P_S}.$$

Since

$$1 - P_S = \frac{e^{-(ax+b_S)}}{1 + e^{-(ax+b_S)}},$$

therefore

$$\Omega_S = \frac{1}{e^{-(ax+b_S)}} = e^{ax+b_S}. \tag{5.27}$$

Similarly, the odds that a nonsmoker from that community will die of lung cancer are

$$\Omega_N = e^{ax+b_N}. \tag{5.28}$$

Thus, if the logistic model is correct, the odds ratio, that is, the ratio of (5.27) to (5.28), becomes simply

$$\omega = \frac{\Omega_S}{\Omega_N} = \frac{e^{ax+b_S}}{e^{ax+b_N}} = e^{(b_S-b_N)}, \tag{5.29}$$

independent of x. The logarithm of the odds ratio is then simply

$$\log(\omega) = b_S - b_N, \tag{5.30}$$

which is also independent of x and, moreover, is the simple difference between the two parameters assumed to distinguish smokers from nonsmokers.

The importance of this result is that, if the odds ratio or its logarithm is found to be stable across many different kinds of populations, then one may reasonably infer that the logistic model is a fair representation of the phenomenon under study. Given this inference, one may predict the value of the odds ratio in a new population and test the difference between the observed and predicted values; one may predict the effects on mortality of controlling the factor represented by x (in our example, an air pollutant); and one may of course predict the effects on mortality of controlling smoking habits.

The representation (5.30) of the logarithm of ω suggests that the logarithm of the sample odds ratio,

$$L = \log(o), \tag{5.31}$$

is an important measure of association. The standard error of L has been studied by Woolf (1955), Haldane (1956), and Gart (1966). A better estimate of log (ω) was found to be

$$L' = \log(o'),\tag{5.32}$$

where o' is defined in (5.20), and a good estimate of its standard error was found to be

$$\text{s.e.}(L') = \sqrt{\frac{1}{n_{11} + .5} + \frac{1}{n_{12} + .5} + \frac{1}{n_{21} + .5} + \frac{1}{n_{22} + .5}}.\tag{5.33}$$

The logarithms in (5.30) to (5.32) must all be to the base e (see Table A.5).

When the logistic model of (5.25) and (5.26) obtains, log (ω) is seen by (5.30) to be completely independent of x. Even if, instead, a model specified by a cumulative normal distribution is assumed, log (ω) is nearly independent of x (Edwards, 1966; Fleiss, 1970). The logistic model is far more manageable for representing rates and proportions than the cumulative normal model, however, and has been so used by Bartlett (1935), Winsor (1948), Dyke and Patterson (1952), Cox (1958, 1970), Grizzle (1961, 1963), and Maxwell and Everitt (1970).

Problem 5.1. The odds Ω_A and $\Omega_{\bar{A}}$ are defined by (5.5) and (5.8). Prove that $\Omega_A = \Omega_{\bar{A}}$ if and only if $P(B \mid A) = P(B \mid \bar{A})$.

Problem 5.2. The odds ratio ω is defined by (5.13). Prove that $\omega > 1$ if and only if $P(B \mid A) > P(B \mid \bar{A})$.

Problem 5.3. The relative risk r is defined by (5.23) and the odds ratio o by (5.14). Prove that r is approximately equal to o if p_{21} is small relative to p_{22} and if p_{11} is small relative to p_{12}. [*Hint.* $p_{2.} = p_{22}(1 + p_{21}/p_{22})$ and $p_{1.} = p_{12}(1 + p_{11}/p_{12})$.]

Problem 5.4. It has long been known that, among first admissions to American public mental hospitals, schizophrenia as diagnosed by the hospital psychiatrists is more prevalent than the affective disorders, whereas the converse is true for British public mental hospitals. A cooperative study between New York and London psychiatrists was designed to determine the extent to which the difference was a function of differences in diagnostic habits. The following data are from a study reported by Cooper et al. (1972).

(*a*) One hundred and forty-five patients in a New York hospital and 145 in a London hospital were selected for study. The New York hospital diagnosed 82 patients as schizophrenic and 24 as affectively ill, whereas the London hospital diagnosed 51 as schizophrenic and 67 as affectively ill. Ignoring the patients given other diagnoses, set up the resulting fourfold table.

The project psychiatrists made diagnoses using a standard set of criteria after conducting standardized interviews with the patients. In New York, the project diagnosed 43 patients as schizophrenic and 53 as affectively ill. In London, the project diagnosed 33 patients as schizophrenic and 85 as affectively ill. Ignoring the patients given other diagnoses, set up the resulting fourfold table.

The results of the standardized interview served as input to a computer program that yields psychiatric diagnoses. In New York, the computer diagnosed 67 patients as schizophrenic and 27 as affectively ill. In London, the computer diagnosed 56 patients as schizophrenic and 37 as affectively ill. Ignoring the patients given other diagnoses, set up the resulting fourfold table.

(b) Three diagnostic contrasts between New York and London are possible: by the hospitals' diagnoses; by the project's diagnoses; and by the computer's diagnoses. Compute, for each of the three sources of diagnoses, the ratio of the odds that a New York patient will be diagnosed schizophrenic rather than affective to the corresponding odds for London.

How do the odds ratios for the project's and computer's diagnoses compare? How do these two compare with the odds ratio for the hospital's diagnoses? By what proportion has the latter odds ratio been reduced by taking an average of the first two?

(c) For each source of diagnosis, all four cell frequencies are large, indicating that the improved estimate (5.20) may not be necessary. Check that, for each source of diagnosis, the estimate of the odds ratio given by (5.20) is only slightly less than the estimate given by (5.14).

REFERENCES

Altham, P. M. E. (1970a). The measurement of association of rows and columns for an r × s contingency table. J. roy. statist. Soc., Ser. B, 32, 63–73.

Altham, P. M. E. (1970b). The measurement of association in a contingency table: Three extensions of the cross-ratios and metric methods. J. roy. statist. soc., Ser. B, 32, 395–407.

Anscombe, F. J. (1956). On estimating binomial response relations. Biometrika, 43, 461–464.

Bartlett, M. S. (1935). Contingency table interactions. J. roy. statist. Soc. Suppl., 2, 248–252.

Berger, A. (1961). On comparing intensities of association between two binary characteristics in two different populations. J. Amer. statist. Assoc., 56, 889–908.

Carroll, J. B. (1961). The nature of the data, or how to choose a correlation coefficient. Psychometrika, 26, 347–372.

Cooper, J. E., Kendell, R. E., Gurland, B. J., Sharpe, L., Copeland, J. R. M. and Simon, R. (1972). Psychiatric diagnosis in New York and London. London: Oxford University Press.

Cornfield, J. (1951). A method of estimating comparative rates from clinical data. Applications to cancer of the lung, breast and cervix. J. natl. Cancer Inst., 11, 1269–1275.

Cox, D. R. (1958). The regression analysis of binary sequences. *J. roy. statist. Soc., Ser. B,* **20**, 215–242.

Cox, D. R. (1970). *Analysis of binary data.* London: Methuen.

Craddock, J. M. and Flood, C. R. (1970). The distribution of the χ^2 statistic in small contingency tables. *Appl. Statist.,* **19**, 173–181.

Dixon, W. J. and Massey, F. J. (1957). *Introduction to statistical analysis, 2nd ed.* New York: McGraw-Hill.

Dyke, G. V. and Patterson, H. D. (1952). Analysis of factorial arrangements when the data are proportions. *Biometrics,* **8**, 1–12.

Edwards, A. W. F. (1963). The measure of association in a 2 × 2 table. *J. roy. statist. Soc., Ser. A,* **126**, 109–114.

Edwards, J. H. (1966). Some taxonomic implications of a curious feature of the bivariate normal surface. *Brit. J. prev. soc. Med.,* **20**, 42–43.

Fisher, R. A. (1954). *Statistical methods for research workers, 12th ed.* Edinburgh: Oliver and Boyd.

Fleiss, J. L. (1970). On the asserted invariance of the odds ratio. *Brit. J. prev. soc. Med.,* **24**, 45–46.

Gart, J. J. (1966). Alternative analyses of contingency tables. *J. roy. statist. Soc., Ser. B,* **28**, 164–179.

Gart, J. J. and Zweifel, J. R. (1967). On the bias of various estimators of the logit and its variance, with application to quantal bioassay. *Biometrika,* **54**, 181–187.

Goodman, L. A. (1964). Simultaneous confidence limits for cross-product ratios in contingency tables. *J. roy. statist. Soc., Ser. B,* **26**, 86–102.

Goodman, L. A. and Kruskal, W. H. (1954). Measures of association for cross classifications. *J. Amer. statist. Assoc.,* **49**, 732–764.

Goodman, L. A. and Kruskal, W. H. (1959). Measures of association for cross classifications. II: Further discussion and references. *J. Amer. statist. Assoc.,* **54**, 123–163.

Goodman, L. A. and Kruskal, W. H. (1963). Measures of association for cross classifications. III: Approximate sampling theory. *J. Amer. statist. Assoc.,* **58**, 310–364.

Grizzle, J. E. (1961). A new method of testing hypotheses and estimating parameters for the logistic model. *Biometrics,* **17**, 372–385.

Grizzle, J. E. (1963). Tests of linear hypotheses when the data are proportions. *Amer. J. public Health,* **53**, 970–976.

Haldane, J. B. S. (1956). The estimation and significance of the logarithm of a ratio of frequencies. *Ann. hum. Genet.,* **20**, 309–311.

Harman, H. H. (1960). *Modern factor analysis.* Chicago: University of Chicago Press.

Lord, F. M. and Novick, M. R. (1968). *Statistical theories of mental test scores.* Reading, Mass.: Addison-Wesley.

Maxwell, A. E. (1961). *Analysing qualitative data.* London: Methuen.

Maxwell, A. E. and Everitt, B. S. (1970). The analysis of categorical data using a transformation. *Brit. J. math. statist. Psychol.,* **23**, 177–187.

Mosteller, F. (1968). Association and estimation in contingency tables. *J. Amer. statist. Assoc.,* **63**, 1–28.

Nunnally, J. (1967). *Psychometric theory.* New York: McGraw-Hill.

Winsor, C. P. (1948). Factorial analysis of a multiple dichotomy. *Hum. Biol.,* **20**, 195–204.

Woolf, B. (1955). On estimating the relation between blood group and disease. *Ann. hum. Genet.*, **19**, 251–253.

Yule, G. U. (1900). On the association of attributes in statistics. *Philos. Trans. roy. Soc. Ser. A*, **194**, 257.

Yule, G. U. (1912). On the methods of measuring the association between two attributes. *J. roy. statist. Soc.*, **75**, 579–642.

CHAPTER 6

Sampling Method II: Prospective and Retrospective Studies

Sampling method II was defined in Section 2.1 as the selection of a sample from each of two populations, a predetermined number n_1 from the first and a predetermined number n_2 from the second. Method II sampling is used in comparative prospective studies—in which one of the two populations is defined by the presence and the second by the absence of a suspected antecedent factor (MacMahon, Pugh, and Ipsen, 1960, Chapter 12)—and is used in comparative retrospective studies—in which one of the two populations is defined by the presence and the second by the absence of the outcome under study (MacMahon, Pugh, and Ipsen, 1960, Chapter 13).

The analysis of data from a comparative prospective study is discussed in Section 6.1, and of data from a comparative retrospective study in Section 6.2. Berkson's and Sheps's criticisms of the odds ratio are presented in Section 6.3, and a comparison of the prospective and retrospective approaches is made in Section 6.4.

6.1. PROSPECTIVE STUDIES

The comparative prospective study (also termed the *cohort*, or *forward-going*, or *follow-up* study) is characterized by the identification of the two study samples on the basis of the presence or absence of the antecedent factor, and by the estimation for both samples of the proportions developing the disease or condition under study.

Consider again the hypothetical association between maternal age—the antecedent factor—and birthweight—the outcome—introduced in Chapter 5. A design fitting the paradigm of a comparative prospective study would be indicated if, for example, the file containing records for mothers aged 20 years or less were kept separate from the file containing records for mothers aged more than 20 years. Suppose that 100 of both kinds of mothers are sampled from the respective lists, and the weights of their offspring ascertained.

The precise outcome is of course subject to chance variation, but let us suppose that the data turn out to be perfectly consistent with those obtained with sampling method I (see Section 5.1). From Table 5.2, the rate of low birthweight specific to mothers aged 20 years or less is estimated to be

$$p(B \mid A) = \frac{p_{11}}{p_{1.}} = \frac{.05}{.25} = .20. \tag{6.1}$$

Thus we would expect to have 20% of the offspring of mothers aged 20 years or less, or 20 infants, weighing 2500 grams or less, and the remaining 80 weighing over 2500 grams.

The rate of low birthweight specific to mothers aged over 20 years is estimated from Table 5.2 to be

$$p(B \mid \bar{A}) = \frac{p_{21}}{p_{2.}} = \frac{.075}{.75} = .10. \tag{6.2}$$

We would therefore expect to have ten of the offspring of mothers aged over 20 years weighing 2500 grams or less, and the remaining 90 weighing over 2500 grams. The expected table is therefore as shown in Table 6.1.

Table 6.1. Association between birthweight and maternal age: prospective study

Maternal Age	Birthweight		Total	Proportion with Low Birthweight
	B	$\bar{B}$		
A	20	80	100 $(= N_A)$	.20 $(= p(B \mid A))$
$\bar{A}$	10	90	100 $(= N_{\bar{A}})$	.10 $(= p(B \mid \bar{A}))$
Total	30	170	200	

The value of χ^2 for these data is

$$\chi^2 = 3.18, \tag{6.3}$$

so that the association fails to reach significance at the .05 level. What is noteworthy is that the total sample sizes of Tables 5.1 and 6.1 are equal and that the frequencies of the two tables are consistent. Nevertheless, the chi square value for the latter table is greater than that for the former. The inference from this comparison holds in general: a prospective study with equal sample sizes yields a more *powerful* chi square test than a cross-sectional study with the same total sample size (see Lehmann, 1959, p. 146).

The odds ratio ω was introduced in Section 5.3 as a measure of association between characteristics A and B. Because of the way the separate odds Ω_A (5.5) and $\Omega_{\bar{A}}$ (5.8) were defined, it is clear that the odds ratio may be estimated from a comparative prospective study as well as from a cross-sectional study. The estimate is

$$o = \frac{p(B \mid A)p(\bar{B} \mid \bar{A})}{p(\bar{B} \mid A)p(B \mid \bar{A})}. \tag{6.4}$$

For the data of Table 6.1, the estimated odds ratio is

$$o = \frac{.20 \times .90}{.80 \times .10} = 2.25, \tag{6.5}$$

precisely equal to the value found in (5.15) for the data from the cross-sectional study.

Formula 5.16 for o as a function of the cross-products of cell frequencies applies to the data from comparative prospective studies as well. For the data of Table 6.1, obviously,

$$o = \frac{20 \times 90}{80 \times 10} = 2.25.$$

The standard error of the odds ratio estimated from a comparative prospective study is approximately

$$\text{s.e.}(o) = o\sqrt{\frac{1}{N_A p(B \mid A)p(\bar{B} \mid A)} + \frac{1}{N_{\bar{A}} p(B \mid \bar{A})p(\bar{B} \mid \bar{A})}}. \tag{6.6}$$

For the data of Table 6.1,

$$\text{s.e.}(o) = 2.25\sqrt{\frac{1}{100 \times .20 \times .80} + \frac{1}{100 \times .10 \times .90}}$$

$$= 0.94. \tag{6.7}$$

An equivalent expression for the standard error in terms of the original frequencies (see Problem 6.1) is

$$\text{s.e.}(o) = o\sqrt{\frac{1}{n_{11}} + \frac{1}{n_{12}} + \frac{1}{n_{21}} + \frac{1}{n_{22}}}, \tag{6.8}$$

identical to expression (5.19).

We found above that the chi square test applied to data from a comparative prospective study with equal sample sizes is more powerful than the chi square test applied to data from a cross-sectional study. A similar phenomenon holds for the *precision* of the estimated odds ratio. Even though the total sample sizes in Tables 5.1 and 6.1 are equal, and even though the association between the two characteristics is the same, nevertheless the odds ratio from the latter table is estimated more precisely [s.e.$(o) = 0.94$, from (6.7)] than the odds ratio from the former [s.e.$(o) = 1.00$, from (5.18)]. A prospective study with equal sample sizes is thus superior in terms of both power and precision to a cross-sectional study with the same total sample size.

The value of the uncorrected chi square statistic (see equation 5.3) for the data of Table 6.1 is

$$\chi^2_u = \frac{200(20 \times 90 - 80 \times 10)^2}{100 \times 100 \times 30 \times 170} = 3.92.$$

It yields a phi coefficient (see equation 5.2) of

$$\varphi = \sqrt{\frac{3.92}{200}} = .14, \tag{6.9}$$

only slightly greater than the value of .13 found in (5.4) for the data of Table 5.1. Recall, however, that the data of Tables 5.1 and 6.1 are perfectly consistent.

6.2. RETROSPECTIVE STUDIES

The comparative retrospective study (also termed the *case-history* study) is characterized by the identification of the two study samples on the basis of the presence or absence of the outcome factor, and by the estimation for both samples of the proportions possessing the antecedent factor under study.

This method might be easily applied to the study of the association between maternal age and birthweight if, for example, the file containing records for infants of low birthweight (less than or equal to 2500 grams) were kept separate from the file containing records for infants of higher birthweight. Suppose that 100 of both kinds of infants are sampled from the respective lists, and the ages of their mothers ascertained.

Let us suppose, as we did in Section 6.1, that the data turn out to be perfectly consistent with those already given. We need the rates $p(A \mid B)$ and $p(A \mid \bar{B})$, that is, the proportions of mothers aged 20 years or less among infants of low and among infants of high birthweight. From Table 5.2 we find that

$$p(A \mid B) = \frac{p_{11}}{p_{.1}} = \frac{.05}{.125} = .40, \tag{6.10}$$

implying that 40% of the mothers of the 100 low-birthweight infants, or 40, should be aged 20 years or less, and the remaining 60, over 20 years. We also find that

$$p(A \mid \bar{B}) = \frac{p_{12}}{p_{.2}} = \frac{.20}{.875} = .23, \tag{6.11}$$

implying that 23% of the mothers of the 100 higher-birthweight infants, or 23, should be aged 20 years or less, and the remaining 77, over 20 years.

Standard practice is to set out the data resulting from sampling method II as in Table 2.6, so that the two study samples appear one above the other and the characteristic determined for each subject is located across the top. This causes something of an anomaly when applied to data from a comparative retrospective study because the characteristic determined for each subject is the suspected antecedent factor and not the outcome characteristic that is hypothesized to follow from it. Miettinen (1970) stressed the necessary shift in thinking required for analyzing data from a comparative retrospective study: that the outcome characteristic follows the antecedent factor in a logical sequence, but precedes it in a retrospective study.

Following, then, the format of Table 2.6, we present the expected data as

Table 6.2. Association between birthweight and maternal age: retrospective study

Birthweight	Maternal Age		Total	Proportion with Low Age
	A	$\bar{A}$		
B	40	60	$100 \ (= N_B)$	$.40 \ (= p(A \mid B))$
$\bar{B}$	23	77	$100 \ (= N_{\bar{B}})$	$.23 \ (= p(A \mid \bar{B}))$
Total	63	137	200	

shown in Table 6.2. The value of χ^2 for these data is

$$\chi^2 = 5.93, \tag{6.12}$$

which indicates that the association is significant at the .05 level.

The gradient in the magnitude of χ^2 from the cross-sectional study ($\chi^2 = 2.58$) to the prospective study ($\chi^2 = 3.18$) to the retrospective study ($\chi^2 = 5.93$) is noteworthy because the three sets of data were all generated according to the same set of underlying rates and because the three total sample sizes were equal. It is true in general that a retrospective study with equal sample sizes yields a more powerful chi square test than a cross-sectional study with the same total sample size. If, in addition, the outcome characteristic is rarer than the antecedent factor (more precisely, if

$$|P(B) - .5| > |P(A) - .5|), \tag{6.13}$$

then the chi square test on the data of a retrospective study with equal sample sizes is more powerful than the chi square test on the data of a prospective study with equal sample sizes (Lehmann, 1959, p. 146).

For the data of Table 6.2, the value of the uncorrected chi square statistic is

$$\chi_u^2 = \frac{200(40 \times 77 - 60 \times 23)^2}{100 \times 100 \times 63 \times 137} = 6.70.$$

The value of the associated phi coefficient is

$$\varphi = \sqrt{\frac{6.70}{200}} = .18, \tag{6.14}$$

nearly 40% higher than the value, $\varphi = .13$, associated with the data of Table 5.1 and nearly 30% higher than the value, $\varphi = .14$, associated with the data of Table 6.1. Now, if a given measure is to be more than a mere uninterpretable index, it should have the property that different investigators studying the same phenomenon should emerge with at least similar estimates even though they studied the phenomenon differently. Since the phi coefficient obviously lacks this property of *invariance*, it should not be used as a measure of association for data from comparative prospective or retrospective studies.

The odds ratio ω, on the other hand, is invariant across the three kinds of studies we are considering. It is defined by

$$\omega = \frac{P(B \mid A)P(\bar{B} \mid \bar{A})}{P(\bar{B} \mid A)P(B \mid \bar{A})}. \tag{6.15}$$

When expressed in this form, ω seems to be estimable only from cross-sectional and comparative prospective studies, because only these two kinds of studies provide estimates of the rates $P(B \mid A)$ and $P(B \mid \bar{A})$. An equivalent expression for ω (see Problem 6.2), however, is

$$\omega = \frac{P(A \mid B)P(\bar{A} \mid \bar{B})}{P(\bar{A} \mid B)P(A \mid \bar{B})}, \tag{6.16}$$

and, when expressed in this form, ω is clearly estimable from comparative retrospective studies, too (see Cornfield, 1956).

The estimate is

$$o = \frac{p(A \mid B)p(\bar{A} \mid \bar{B})}{p(\bar{A} \mid B)p(A \mid \bar{B})}. \tag{6.17}$$

For the data of Table 6.2, the estimated odds ratio is

$$o = \frac{.40 \times .77}{.60 \times .23} = 2.23, \tag{6.18}$$

equal except for rounding errors to the value $o = 2.25$ found previously. Formula 5.16 continues to apply as well.

The standard error of the odds ratio estimated from a comparative retrospective study is approximately

$$\text{s.e.}(o) = o\sqrt{\frac{1}{N_B p(A \mid B)p(\bar{A} \mid B)} + \frac{1}{N_{\bar{B}} p(A \mid \bar{B})p(\bar{A} \mid \bar{B})}}. \quad (6.19)$$

Expressions 5.19 and 6.8 for the standard error as a function of the original frequencies are also valid for the data of a comparative retrospective study.

For the data of Table 6.2,

$$\text{s.e.}(o) = 2.23\sqrt{\frac{1}{100 \times .40 \times .60} + \frac{1}{100 \times .23 \times .77}}$$

$$= 2.23\sqrt{\frac{1}{40} + \frac{1}{60} + \frac{1}{23} + \frac{1}{77}}$$

$$= 0.70. \quad (6.20)$$

The gradient noted above in the magnitude of χ^2 is matched by a gradient in the precision with which the odds ratio is estimated. For the cross-sectional study, s.e.$(o) = 1.00$ (5.18); for the comparative prospective study, s.e.$(o) = 0.94$ (6.7); and for the comparative retrospective study, s.e.$(o) = 0.70$ (6.20). Thus, according to the criterion of precision as well as the criterion of power, a comparative retrospective study is superior to both the cross-sectional and the comparative prospective methods of study.

6.3. CRITICISMS OF THE ODDS RATIO

It was pointed out in the paragraph following (6.15) that the two rates $P(B \mid A)$ and $P(B \mid \bar{A})$ are estimable only from cross-sectional and prospective studies (we continue to let A denote the presence of the antecedent factor and B the undesirable outcome). Only when association is measured by the odds ratio, which is a function of the *ratio* of these rates, does the retrospective study provide data comparable to the data from the other two kinds of studies.

Berkson (1958), however, has strongly criticized taking the ratio of rates as a measure of association, pointing out that the level of the rates is lost. Thus a tenfold increase over a rate of one per million would be considered equivalent to a tenfold increase over a rate of one per thousand, even though the latter increase is far more serious than the former. Berkson maintains that the simple difference between two rates is the proper measure of the practical magnitude, in terms of public health importance, of an association.

Data from Table 26 of *Smoking and Health* (1964, p. 110) will illustrate Berkson's point. Table 6.3 gives the approximate death rates per 100,000 person-years for smokers and for nonsmokers of cigarettes.

Table 6.3. Mortality rates per 100,000 *person-years from lung cancer and coronary artery disease for smokers and nonsmokers of cigarettes*

	Smokers	Nonsmokers	o	Difference
Cancer of the Lung	48.33	4.49	10.8	43.84
Coronary Artery Disease	294.67	169.54	1.7	125.13

If we compared only the odds ratios, o, we would conclude that cigarette smoking has a greater effect on lung cancer than on coronary artery disease. It was this conclusion, from a number of studies, that Berkson felt was unwarranted. He contended, quite correctly, that the odds ratio throws away all information on the number of deaths due to either cause. Berkson went farther, however, and maintained that it is *only* the excess in mortality that permits a valid assessment of the effect of smoking on a cause of death: ". . . of course, from a strictly practical viewpoint, it is only the total number of increased deaths that matters [1958, p. 30]."

He maintained that the effect of smoking is greater on that cause of death with the greater number of excess deaths. Thus, since smoking is associated with an excess mortality of over 120 per 100,000 person-years from coronary artery disease and with an excess mortality of under 50 per 100,000 person-years from lung cancer, Berkson concluded that the association is stronger for coronary artery disease than for lung cancer.

Sheps (1958, 1961) has proposed a simple and elegant modification of Berkson's index. Let p_c denote the mortality rate (in general, the rate at which an untoward event occurs) in a control sample, and p_s the corresponding rate in a study sample presumed to be at higher risk. Thus, by assumption, $p_s > p_c$.

Sheps contends that the excess risk associated with being in the study group, say p_e, can operate only on those individuals who would not have had the event occurring to them otherwise. Thus she postulates the model

$$p_s = p_c + p_e(1 - p_c); \tag{6.21}$$

that is, the rate in the study group, p_s, is the sum of the rate in the control group, p_c, and of the excess risk, p_e, applied to those who would not otherwise have had the event, $1 - p_c$. Sheps suggests that p_e be used as a measure of added or excess risk. Because, clearly,

$$p_e = \frac{p_s - p_c}{1 - p_c}, \tag{6.22}$$

p_e may also be called the *relative difference.*

This differs from Berkson's index, $p_s - p_c$, only in that the difference is divided by $1 - p_c$, the proportion of the people actually at added risk. If p_c

is small, then Sheps's and Berkson's indexes are close. Thus, for the data of Table 6.3, the excess mortality due to cancer of the lung is $p_e = 43.84/(100,000 - 4.49) = 43.84$ additional deaths attributable to cigarette smoking per 100,000 person-years saved by not having smoked. For coronary artery disease, the excess mortality is $p_e = 125.13/(100,000 - 169.54) = 125.34$ deaths per 100,000 person-years saved by not having smoked.

If research into the etiology of disease were concerned *solely* with public health issues, then Berkson's simple difference or Sheps's relative difference would be the only valid measures of the association between the antecedent factor and the outcome event. Because retrospective studies are incapable of providing estimates of either measure, such studies are necessarily useless from the point of view of public health. As Cornfield et al. (1959) and Greenberg (1969) have pointed out, however, etiological research is also concerned with the search for regularities in many sets of data, with the development of models of disease causation and distribution, and with the generation of hypotheses by one set of data that can be tested on other sets.

Given these concerns, that measure of association is best which is suggested by a mathematical model, which remains valid under alternative models, which is capable of assuming predicted values for certain kinds of populations and can thus serve as the basis of a test of a hypothesis, and which is invariant under different methods of studying association. Because the odds ratio, or such a function of it as its logarithm, comes closest to providing such a measure (Cox, 1970, pp. 20–21), and because the odds ratio is estimable from a retrospective study, retrospective studies are eminently valid from the more general point of view of the advancement of knowledge. Peacock (1971) warns, however, against the uncritical assumption that the odds ratio should be constant across different kinds of populations.

That Sheps's measure, as does Berkson's, lacks the regularity observed with the odds ratio is seen from the data of Table 6.4. It gives overall death rates for smokers and nonsmokers of varying ages, taken from the graph on page 88 of *Smoking and Health* (1964). In that graph two rates appear for

Table 6.4. Death rates from all causes per 100,000 by age and smoking

Age Interval	Smokers	Nonsmokers	o	p_e per 100,000
45–49	580	270	2.2	310
50–54	1050	440	2.4	610
55–59	1600	850	1.9	750
60–64	2500	1500	1.7	1000
65–69	3700	2000	1.9	1700
70–74	5300	3000	1.8	2400
75–79	9200	4800	2.0	4600

the nonsmokers aged 75–79. The value that seemed the more reasonable was used.

As is so often the case, one comes to different conclusions depending on which measure he chooses. Looking at p_e, the conclusion is that the effect of smoking steadily increases with increasing age. In terms of lives lost, among those not otherwise expected to die, this conclusion is correct. The increase in p_e, however, is so erratic that no precise mathematical extrapolation or even interpolation seems possible.

Looking at o, on the other hand, the conclusion is that the effect of smoking on overall mortality is essentially constant across all the tabulated ages. The epidemiological importance of this conclusion is that one may validly predict that the odds ratio would be approximately 2.0 for specific ages between 45 and 79, for ages below 45, and for ages above 79. If the observed odds ratio for any such interpolated or extrapolated age departed appreciably from 2.0, further research would clearly be indicated.

6.4. THE RETROSPECTIVE APPROACH VERSUS THE PROSPECTIVE APPROACH

If a scientist accepts the argument of Section 6.3 that the odds ratio and thus the retrospective study are inherently valid, he must still bear in mind that retrospective studies are subject to more sources of error than prospective studies. Hammond (1958), for example, has cited the bias that may arise because historical data are obtained only after subjects become ill, and frequently only after they are diagnosed. A patient's knowledge that he has a certain disease might easily affect his recollection, intentionally or unintentionally, of which factors preceded his illness.

Another difficulty pointed out by Hammond is in finding an adequate control series for the sample of patients: one must, after all, find groups of subjects who are like the cases in all respects save for having the disease. Mantel and Haenszel (1959) cite these and other deficiencies in the retrospective approach. When, for example, the subjects having the disease are found in hospitals or clinics, inferences from retrospective studies may be subject to the kind of bias illustrated in Section 1.3. Specifically, the antecedent factors may appear to be associated with the disease, but might in reality be more associated with admission to a treatment facility.

This does not imply that only the retrospective approach, and not the prospective, is open to bias. Similar biases have been shown to operate in prospective studies as well (Mainland and Herrera, 1956; Yerushalmy and Palmer, 1959; Mantel and Haenszel, 1959). For example, the bias possible in those retrospective studies that require hospitalized patients to be evaluated is matched by the potential bias in those prospective studies that require volunteers to be followed up.

What does seem to be true, however, is that a greater degree of ingenuity is needed for the proper design of a retrospective study than for the design of a prospective study. Thus Levin (1954) controlled for the first bias cited above—that due to a patient's report possibly being influenced by his knowledge that he has the disease being studied—by questioning all subjects prior to the final diagnosis. By this means, the possible bias due to the examiner's applying different standards to the responses of cases and controls is also controlled. Rimmer and Chambers (1969) suggested another means of control. They found greater accuracy in the recollections of relatives than in those of the patients themselves.

As a means of reducing the bias possible in contrasting cases with only one control series, Doll and Hill (1952) studied two control groups. One was a sample of hospitalized patients with other diseases than the one studied, and the second was a sample from the community. Matching (see Chapter 8) is another device for reducing bias. The validity of the retrospective approach can only increase as investigators learn which kinds of information can be accurately recalled by a subject and which cannot. For example, Gray (1955) and Klemetti and Saxén (1967) have shown that the occurrence or non-occurrence of a past event can be recalled accurately, but not the precise time when the event occurred.

Some other means of reducing errors are presented in Chapter 12. To the extent that bias can be controlled, the following points made by Mantel and Haenszel argue strongly for the retrospective approach:

Among the desirable attributes of the retrospective study is the ability to yield results from presently collectible data . . . . The retrospective approach is also adapted to the limited resources of an individual investigator . . . . For especially rare diseases a retrospective study may be the only feasible approach . . . . In the absence of important biases in the study setting, the retrospective method could be regarded, according to sound statistical theory, as the study method of choice [1959, p. 720].

Problem 6.1. Prove the equality of expressions 6.6 and 6.8 for the standard error of the odds ratio. [*Hint.* $p(B \mid A) = n_{11}/N_A$; $p(\bar{B} \mid A) = n_{12}/N_A$; $p(B \mid \bar{A}) = n_{21}/N_{\bar{A}}$; and $p(\bar{B} \mid \bar{A}) = n_{22}/N_{\bar{A}}$. Also, $N_A = n_{11} + n_{12}$ and $N_{\bar{A}} = n_{21} + n_{22}$.]

Problem 6.2. Prove the equality of expressions 6.15 and 6.16 for the odds ratio. [*Hint.* Use the definitions of Section 1.1 to replace all conditional probabilities in (6.15) by joint probabilities. The probabilities $P(A)$ and $P(\bar{A})$ are seen to cancel out. Multiply and divide by $P(B)$ and by $P(\bar{B})$, and use the definition of conditional probabilities to arrive at (6.16).]

Problem 6.3. The phi coefficient (5.2) is a valid measure of association only for method I (naturalistic or cross-sectional) sampling. Phi coefficients applied to data from method II (prospective or retrospective) studies are not at all comparable to those applied to data from method I studies.

Even more is true. When two studies are both conducted according to either the prospective or retrospective approaches, but with proportionately different allocations of the total sample, the phi coefficient for one will not in general be comparable to that for the other.

(a) In a retrospective study of factors associated with cancer of the oral cavity, Wynder, Navarrette, Arostegui, and Llambes (1958) studied 34 women with cancer of the oral cavity and 214 women, matched by age, with nonmalignant conditions. Twenty-four percent of the cancer cases, as opposed to 66% of the controls, were nonsmokers. Set up the resulting fourfold table and calculate uncorrected chi square and the associated phi coefficient.

(b) Suppose that Wynder et al. had studied, instead, 214 cancer cases and 34 controls. Assuming the same proportions of nonsmokers as above, set up the expected fourfold table and calculate uncorrected chi square and the associated phi. How do the phi coefficients compare?

(c) Suppose now, that 124 of both kinds of women had been studied, and assume the same proportions of nonsmokers as above. Set up the resulting expected fourfold table and calculate uncorrected chi square and the associated phi. How does this phi coefficient compare with those in (a) and (b)? What is the relative difference between the phi coefficient of (a) and that of (c)? What would you conclude about the comparability of phi coefficients in retrospective studies with varying allocations of a total sample?

Problem 6.4. Two criteria are suggested in this chapter for comparing the cross-sectional, prospective, and retrospective approaches: the values of chi square to be expected if total sample sizes were equal, and the approximate standard errors of the odds ratio to be expected if total sample sizes were equal. Another criterion for comparison is the total sample size necessary for the standard error to assume some specified value.

The data for the following questions are those employed throughout Chapters 5 and 6. Suppose that the correct values of $o = 2.25$ and of all the necessary proportions are known.

(a) The approximate standard error of o with cross-sectional sampling is given by (5.17). What value of n is needed to give a standard error of 0.50?

(b) The approximate standard error of o with prospective sampling is given by (6.6). Let $N_A + N_{\bar{A}}$, the total sample size, be denoted N_P, and suppose for simplicity that $N_A = N_{\bar{A}} = N_P/2$. What value of N_P is needed to give a standard error of 0.50? What is the percentage reduction from $n_{..}$ to N_P?

(c) The approximate standard error of o with retrospective sampling is given by (6.19). Let $N_B + N_{\bar{B}}$, the total sample size, be denoted N_R, and suppose for simplicity that $N_B = N_{\bar{B}} = N_R/2$. What value of N_R is needed to give a standard error of 0.50? What is the percentage reduction from $n_{..}$ to N_R? From N_P to N_R?

(d) Is the reduction in (b) of much practical (e.g., monetary) importance? How about the reductions in (c)?

REFERENCES

Berkson, J. (1958). Smoking and lung cancer: Some observations on two recent reports. *J. Amer. statist. Assoc.*, **53**, 28–38.

Cornfield, J. (1956). A statistical problem arising from retrospective studies. Pp. 135–148 in J. Neyman (Ed.). *Proceedings of the third Berkeley symposium on mathematical statistics and probability*, Vol. IV. Berkeley: University of California Press.

Cornfield, J., Haenszel, W., Hammond, E. C., Lilienfeld, A. M., Shimkin, M. B. and Wynder, E. L. (1959). Smoking and lung cancer: Recent evidence and a discussion of some questions. *J. natl. Cancer Inst.*, **22**, 173–203.

Cox, D. R. (1970). *Analysis of binary data.* London: Methuen.

Doll, R. and Hill, A. B. (1952). A study of the etiology of carcinoma of the lung. *Brit. med. J.*, **2**, 1271–1286.

Gray, P. G. (1955). The memory factor in social surveys. *J. Amer. statist. Assoc.*, **50**, 344–363.

Greenberg, B. G. (1969). Problems of statistical inference in health with special reference to the cigarette smoking and lung cancer controversy. *J. Amer. statist. Assoc.*, **64**, 739–758.

Hammond, E. C. (1958). Smoking and death rates: A riddle in cause and effect. *Amer. Scientist*, **46**, 331–354.

Klemetti, A. and Saxén, L. (1967). Prospective versus retrospective approach in the search for environmental causes of malformations. *Amer. J. public Health*, **57**, 2071–2075.

Lehmann, E. L. (1959). *Testing statistical hypotheses.* New York: Wiley.

Levin, M. L. (1954). Etiology of lung cancer: Present status. *N.Y. State J. Med.*, **54**, 769–777.

MacMahon, B., Pugh, T. F. and Ipsen, J. (1960). *Epidemiologic methods.* Boston: Little, Brown.

Mainland, D. and Herrera, L. (1956). The risk of biased selection in forward-going surveys with nonprofessional interviewers. *J. chronic Dis.*, **4**, 240–244.

Mantel, N. and Haenszel, W. (1959). Statistical aspects of the analysis of data from retrospective studies of disease. *J. natl. Cancer Inst.*, **22**, 719–748.

Miettinen, O. S. (1970). Matching and design efficiency in retrospective studies. *Amer. J. Epidemiol.*, **91**, 111–118.

Peacock, P. B. (1971). The noncomparability of relative risks from different studies. *Biometrics*, **27**, 903–907.

Rimmer, J. and Chambers, D. S. (1969). Alcoholism: Methodological considerations in the study of family illness. *Amer. J. Orthopsychiatr.*, **39**, 760–768.

Sheps, M. C. (1958). Shall we count the living or the dead? *New Eng. J. Med.*, **259**, 1210–1214.

Sheps, M. C. (1961). Marriage and mortality. *Amer. J. public Health*, **51**, 547–555.

Smoking and Health (1964). Report of the Advisory Committee to the Surgeon General of the Public Health Service. Princeton: Van Nostrand.

Wynder, E. L., Navarrette, A., Arostegui, G. E. and Llambes, J. L. (1958). Study of environmental factors in cancer of the respiratory tract in Cuba. *J. natl. Cancer Inst.*, **20**, 665–673.

Yerushalmy, J. and Palmer, C. E. (1959). On the methodology of investigations of etiologic factors in chronic diseases. *J. chronic Dis.*, **10**, 27–40.

CHAPTER 7

Sampling Method III: Controlled Comparative Trials

Sampling method III is exemplified by the comparative clinical trial in which treatments are assigned to subjects at random. The philosophy of the controlled clinical trial is discussed by Hill (1962, Chapters 1 to 3), and solutions to some practical problems arising in the execution of a clinical trial are offered by Mainland (1960). Ethical issues are discussed by Fox (1959). In this brief chapter we only consider the case where the outcome is a yes-no variable such as recovery-no recovery.

The briefness of this chapter is not a reflection of any possible lack of importance of controlled trials. On the contrary, the uniqueness of comparative clinical trials as studies aimed as much at the testing of scientific hypotheses as at the alleviation of symptoms of the subjects being studied has been reflected in a change in thinking about their proper design. Whereas the classic design calls for assigning the different treatments to equal numbers of patients, the newer ones call for assigning to an increasing proportion of patients that treatment which, on the basis of the accumulated data, appears to be superior.

Because the newer designs are still in the process of development, and because they call for complex planning, only the classic design—even though it may soon be out of date—is considered. The interested reader is referred to Anscombe (1963), Colton (1963), Cornfield, Halperin, and Greenhouse (1969), Zelen (1969), and Canner (1970) for some suggested means of balancing the statistical requirement of equal sample sizes with the ethical requirement of applying the superior treatment to as many patients as possible as quickly as possible.

Section 7.1 describes the analysis of data from a simple comparative trial, and Section 7.2 discusses the crossover design.

7.1. THE SIMPLE COMPARATIVE TRIAL

Suppose that the data of Table 7.1 resulted from a trial in which one

Table 7.1.   Hypothetical data from a comparative clinical trial

	Number of Patients	Proportion Improved
Treatment 1	80 ($= n_1$)	.60 ($= p_1$)
Treatment 2	70 ($= n_2$)	.80 ($= p_2$)
Overall	150 ($= n$)	.69 ($= \bar{p}$)

treatment was applied to a sample of $n_1 = 80$ subjects randomly selected from a total of $n = 150$ and the other treatment was applied to the remaining $n_2 = 70$ subjects.

The statistical significance of the difference between the two improvement rates is tested using the statistic given in (2.5). For the data of Table 7.1, the value is

$$z = \frac{|.80 - .60| - \frac{1}{2}(\frac{1}{80} + \frac{1}{70})}{\sqrt{.69 \times .31(\frac{1}{80} + \frac{1}{70})}} = 2.47, \tag{7.1}$$

indicating that the difference is significant at the .05 level.

The simple difference between the two improvement rates,

$$d = p_2 - p_1, \tag{7.2}$$

is the measure most frequently used to describe the differential effectiveness of the second treatment over the first. The approximate standard error of d is

$$\text{s.e.}(d) = \sqrt{\frac{p_1 q_1}{n_1} + \frac{p_2 q_2}{n_2}}. \tag{7.3}$$

For the data of Table 7.1, the simple difference is

$$d = .80 - .60 = .20, \tag{7.4}$$

implying that, among every 100 patients given the first treatment, an additional 20 would have been expected to improve had they been given the second treatment. The estimated standard error of d is

$$\text{s.e.}(d) = \sqrt{\frac{.60 \times .40}{80} + \frac{.80 \times .20}{70}}$$
$$= .07. \tag{7.5}$$

One can occasionally assume that the two treatments are such that any patient who responds to the first treatment is also expected to respond to the second. This assumption may be tenable if the first treatment is an inert placebo, or if the first treatment is an active drug and the second is that drug plus another compound or that drug at a greater dosage level. A consequence of the assumption is that any greater effectiveness of the second treatment can only be manifested on subjects who were refractory to the first (see Sheps, 1958, for further examples and discussion).

Let P_1 denote the proportion improving in the population of patients given the first treatment and P_2 the proportion improving in the population given the second. Let f denote the proportion of patients, among those failing to respond to the first treatment, who would be expected to respond to the second. It is then assumed that

$$P_2 = P_1 + f(1 - P_1), \tag{7.6}$$

that is, the improvement rate under the second treatment is equal to that under the first plus an added improvement rate which applies only to patients who fail to improve under the first treatment. The value of f is clearly

$$f = \frac{P_2 - P_1}{1 - P_1}, \tag{7.7}$$

so that f may be termed the *relative difference*.

Because the sample proportions p_1 and p_2 are estimates of the corresponding population proportions, an estimate of the relative difference is

$$p_e = \frac{p_2 - p_1}{1 - p_1}. \tag{7.8}$$

Its standard error (see Sheps, 1959) is approximately

$$\text{s.e.}(p_e) = \frac{1}{q_1} \sqrt{\frac{p_2 q_2}{n_2} + (1 - p_e)^2 \frac{p_1 q_1}{n_1}}. \tag{7.9}$$

For the data of Table 7.1, the relative difference is

$$p_e = \frac{.80 - .60}{1 - .60} = .50, \tag{7.10}$$

implying that, of every 100 patients who fail to improve under the first treatment, 50 would be expected to improve under the second. The estimated standard error of p_e is

$$\text{s.e.}(p_e) = \frac{1}{.40} \sqrt{\frac{.80 \times .20}{70} + (1 - .50)^2 \frac{.60 \times .40}{80}}$$

$$= .14. \tag{7.11}$$

7.2. THE TWO-PERIOD CROSSOVER DESIGN

The next chapter presents means for analyzing data from, *inter alia*, a controlled trial in which patients are first matched on characteristics associated with the outcome and then randomly assigned the treatments. An extreme example of matching is when, as in a *crossover design*, each patient serves as his own control, that is, when each patient receives each treatment.

Half of the sample of patients is randomly selected to be given the two treatments in one order, and the other half to be given the treatments in the reverse order. A number of factors must be guarded against in analyzing the data from such studies, however.

Meier, Free, and Jackson (1958) have shown that the order in which the treatments are given may affect the response. A test that is valid when order effects are present has been described by Gart (1969). Another factor to be guarded against is the possibility that a treatment's effectiveness is long-lasting, and hence may affect the response to the treatment given after it.

When this so-called *carry-over effect* operates, and when it is unequal for the two treatments, Grizzle (1965) has shown that, for comparing their effectiveness, only the data from the first period may be used. Specifically, the responses by the subjects given one of the treatments first must be compared with the responses by the subjects given the other treatment first. The responses to the treatments given second shed light on the carry-over effects, but might just as well not have been determined if the simple effectiveness of the treatments is all that is of interest.

Differential carry-over effects may be eliminated by interposing a long dry-out period between the termination of the treatment given first and the beginning of the treatment given second. The longer the dry-out period, however, the greater the chances that patients drop out of the trial.

Crossover designs are safe when the treatments are short-acting. When the possibility exists that they are long-acting, the crossover design is to be avoided.

Problem 7.1. Suppose that the two treatments contrasted in Table 7.1 are compared in a second hospital, with the following results.

	Number of Patients	Proportion Improved
Treatment 1	100	.35
Treatment 2	100	.75
Overall	200	.55

(*a*) Is the difference between the improvement rates significant in the second hospital?

(b) What is the simple difference (7.2) between these two improvement rates? What is its standard error (7.3)? Is the difference found in the second hospital significantly different from that found in the first? (*Hint.* Denote the difference and its standard error found in (7.4) and (7.5) by d_1 and s.e.(d_1), and denote the corresponding statistics just calculated by d_2 and s.e.(d_2). Refer the value of

$$z = \frac{d_2 - d_1}{\sqrt{(\text{s.e.}(d_1))^2 + (\text{s.e.}(d_2))^2}}$$

to Table A.2 of the normal distribution.)

(c) What is the relative difference (7.8) between these two improvement rates? What is its standard error (7.9)? Is the relative difference found in the second hospital significantly different from that found in the first? (*Hint.* Refer the value of

$$z = \frac{p_{e(2)} - p_{e(1)}}{\sqrt{(\text{s.e.}(p_{e(1)}))^2 + (\text{s.e.}(p_{e(2)}))^2}}$$

to Table A.2.)

REFERENCES

Anscombe, F. J. (1963). Sequential medical trials. *J. Amer. statist. Assoc.*, **58**, 365–384.

Canner, P. L. (1970). Selecting one of two treatments when the responses are dichotomous. *J. Amer. statist. Assoc.*, **65**, 293–306.

Colton, T. (1963). A model for selecting one of two medical treatments. *J. Amer. statist. Assoc.*, **58**, 388–401.

Cornfield, J., Halperin, M. and Greenhouse, S. W. (1969). An adaptive procedure for sequential clinical trials. *J. Amer. statist. Assoc.*, **64**, 759–770.

Fox, T. F. (1959). The ethics of clinical trials. Pp. 222–229 in D. R. Laurence (Ed.). *Quantitative methods in human pharmacology and therapeutics.* New York: Pergamon Press.

Gart, J. J. (1969). An exact test for comparing matched proportions in crossover designs. *Biometrika*, **56**, 75–80.

Grizzle, J. E. (1965). The two-period change-over design and its use in clinical trials. *Biometrics*, **21**, 467–480.

Hill, A. B. (1962). *Statistical methods in clinical and preventive medicine.* New York: Oxford University Press.

Mainland, D. (1960). The clinical trial—some difficulties and suggestions. *J. chronic Dis.*, **11**, 484–496.

Meier, P., Free, S. M. and Jackson, G. L. (1958). Reconsideration of methodology in studies of pain relief. *Biometrics*, **14**, 330–342.

Sheps, M. C. (1958). Shall we count the living or the dead? *New Engl. J. Med.*, **259**, 1210–1214.

Sheps, M. C. (1959). An examination of some methods of comparing several rates or proportions. *Biometrics*, **15**, 87–97.

Zelen, M. (1969). Play the winner rule and the controlled clinical trial. *J. Amer. statist. Assoc.*, **64**, 131–146.

CHAPTER 8

The Analysis of Data from Matched Samples

A device often employed in controlled trials (sampling method III) is to match subjects on the basis of characteristics that are associated with the response being studied, and to randomize the treatment assignments independently within each matched group. Matched pairs of subjects are used for comparing two treatments, matched triples for comparing three treatments, and in general matched *m-tuples* for comparing *m* treatments. The purpose of matching in controlled trials is to increase the precision of the comparisons among the treatments (Hill, 1962, p. 21).

Matching is also frequently employed in comparative prospective and retrospective studies (sampling method II), but more for increasing the validity of the inferences by controlling for *confounding* factors than for increasing precision (see Bross, 1969, and Miettinen, 1970a, for a debate on this point). Age and sex, for example, are possible confounding factors in the study of the association between cigarette smoking and lung cancer, because age and sex are associated both with smoking and with the risk of lung cancer. In a retrospective study, therefore, these factors might be controlled by matching each case of lung cancer with a control subject of the same sex and of a similar age. Because the cases and controls would then be similar on sex and age, any difference between the two samples would have to be attributable to other factors. Section 10.5 presents another device for the control of confounding factors.

Sampling method I does not lend itself to matching.

Section 8.1 is devoted to the analysis of data from matched pairs when only a dichotomous (i.e., yes-no) outcome is of interest, and Section 8.2 to the analysis of data from matched pairs when more than a dichotomous outcome is of interest. Section 8.3 considers the analysis of data resulting from the study of cases matched with multiple controls when the controls all form a single sample. Section 8.4 considers the analysis of such data when each control represents a separate sample. Some comments on the advantages and disadvantages of matching are made in Section 8.5.

8.1. MATCHED PAIRS: DICHOTOMOUS OUTCOME

In Epidemiology. Suppose that a retrospective study has been conducted in which each case has been matched with a single control and in which the relative frequency of an antecedent factor among the cases is to be compared with that among the controls. Because of the matching of cases with controls, the proper unit of analysis is the matched pair rather than the individual subject. Table 8.1 gives the appropriate means for presenting the resulting data.

Table 8.1.   Data on two outcomes from matched pairs

| | Controls | | |
Cases	Factor Present	Factor Absent	Total
Factor present	a	b	$a + b$
Factor absent	c	d	$c + d$
Total	$a + c$	$b + d$	n

Each frequency in Table 8.1 represents a number of *pairs*. Thus there were n pairs studied in all. Of these, a were such that both members (the case and his matched control) had the antecedent factor; b were such that the case had the factor but the control did not; c were such that the control had the factor but the case did not; and d were such that neither member had the factor.

The proportion of controls who had the factor is

$$p_1 = \frac{a + c}{n},$$

and the proportion of cases who had the factor is

$$p_2 = \frac{a + b}{n}.$$

The number of pairs in which both the case and his matched control had the factor, a, clearly does not affect the difference between the two proportions,

$$p_2 - p_1 = \frac{b - c}{n}. \tag{8.1}$$

As McNemar (1947) has shown, neither a nor d, the numbers of pairs both of whose members were similar with respect to the antecedent factor, contributes explicitly to the standard error of the difference,

$$\text{s.e.}(p_2 - p_1) = \frac{\sqrt{b + c}}{n}. \tag{8.2}$$

The square of the ratio of (8.1) to (8.2) may, with a correction for continuity, be used to test for the statistical significance of the difference between p_1 and p_2. The correction, due to Edwards (1948), yields the statistic

$$\chi^2 = \left\{ \frac{|p_2 - p_1| - 1/n}{\text{s.e.}(p_2 - p_1)} \right\}^2 = \frac{(|b - c| - 1)^2}{b + c}. \tag{8.3}$$

The value of χ^2 may be referred to tables of chi square with one degree of freedom (see McNemar, 1947; Mosteller, 1952; and Stuart, 1957). If χ^2 is large, the inference can be made that the cases and controls differ in the proportion having the antecedent factor. It is noteworthy that only the pairs in which the members differ in the antecedent factor contribute to the test statistic. The power of this test has been studied by Miettinen (1968) and by Bennett and Underwood (1970).

The test based on (8.3), termed *McNemar's test*, is illustrated on the hypothetical data of Table 8.2.

**Table 8.2.   Hypothetical data to illustrate McNemar's
test**

	Controls		
Cases	Factor Present	Factor Absent	Total
Factor present	15	20	35
Factor absent	5	60	65
Total	20	80	100

The proportion of controls having the factor is $p_1 = 20/100 = .20$, and the proportion of cases having the factor is $p_2 = 35/100 = .35$. The standard error of the difference [see (8.2)] is s.e.$(p_2 - p_1) = \sqrt{20 + 5}/100 = 5/100 = .05$, and the test statistic [see (8.3)] has the value

$$\chi^2 = \left(\frac{|.35 - .20| - .01}{.05} \right)^2 = \left(\frac{.14}{.05} \right)^2$$

$$= 7.84,$$

equal to the value obtained by comparing the numbers of pairs whose members differed on the factor,

$$\chi^2 = \frac{(|20 - 5| - 1)^2}{20 + 5} = \frac{196}{25} = 7.84.$$

Since χ^2 exceeds 6.63, the value needed for significance at the .01 level, the conclusion can be drawn that the cases and controls differ in the presence of the antecedent factor.

As pointed out in Chapters 5 and 6, the *odds ratio* (the odds of the disease when the factor is present relative to the odds when the factor is absent) is an important measure of the degree of association between the antecedent factor and the disease. Mantel and Haenszel (1959) and Cornfield and Haenszel (1960) have investigated the proper method for estimating the odds ratio when matched pairs have been studied. When the data are arrayed as in Table 8.1, the estimate is simply

$$o = \frac{b}{c}, \tag{8.4}$$

and its approximate standard error is estimated by

$$\text{s.e.}(o) = (1 + o)\sqrt{\frac{o}{b + c}}. \tag{8.5}$$

For the frequencies of Table 8.2, the estimated odds ratio is $o = 20/5 = 4.0$ and its estimated standard error is

$$\text{s.e.}(o) = (1 + 4)\sqrt{\frac{4}{20 + 5}} = 2.0.$$

In a Controlled Trial. The analysis of the fourfold table resulting from matched pairs has been presented in the context of a comparative retrospective study. The analysis involving McNemar's test and the estimation of the odds ratio may also be applied to a comparative prospective study with matched pairs. In the analysis of a controlled trial with matched pairs, however, the finding of a significant difference by McNemar's test is often better followed up by the estimation of the *relative difference*.

Table 8.3 presents the proper means for presenting the data from such a trial, in which we suppose that a new treatment was compared with a standard. As was the case for Table 8.1, each frequency represents a number of pairs.

Table 8.3. Data from a controlled trial with matched pairs

New Treatment	Standard Treatment		Total
	Recovered	Not Recovered	
Recovered	a	b	$a + b$
Not recovered	c	d	$c + d$
Total	$a + c$	$b + d$	n

The proportion of cases who recover under the standard treatment is

$$p_1 = \frac{a + c}{n}$$

and the proportion who recover under the new treatment is

$$p_2 = \frac{a + b}{n}.$$

Under the assumption that the new treatment can benefit only those patients who fail to improve under the standard treatment, the relative value of the new treatment may be estimated by the relative difference,

$$p_e = \frac{p_2 - p_1}{1 - p_1} = \frac{b - c}{b + d}. \tag{8.6}$$

The approximate standard error of the relative difference may be estimated by

$$\text{s.e.}(p_e) = \frac{1}{(b + d)^2} \sqrt{(b + c + d)(bc + bd + cd) - bcd}. \tag{8.7}$$

Note that a, the number of pairs both of whose members recovered, contributes neither to the estimation of the relative difference nor to the estimation of its standard error.

Table 8.4 presents some hypothetical data. Of the patients who were given

**Table 8.4. Hypothetical data from a controlled trial
with matched pairs**

	Standard Treatment		
New Treatment	Recovered	Not Recovered	Total
Recovered	40	25	65
Not recovered	10	0	10
Total	50	25	75

the standard treatment, the proportion who recovered was $p_1 = 50/75 = .67$. Of those who were given the new treatment, the proportion who recovered was $p_2 = 65/75 = .87$. The value of McNemar's chi square statistic (8.3) for assessing the significance of the difference between these two proportions is

$$\chi^2 = \frac{(|25 - 10| - 1)^2}{25 + 10} = 5.60.$$

The difference is therefore statistically significant at the .05 level.

The value of the relative difference (8.6) is

$$p_e = \frac{25 - 10}{25} = .60,$$

which means that, of every 100 patients who fail to recover under the standard treatment, 60 might be expected to recover under the new treatment. The estimated standard error of the relative difference (8.7) is

$$\text{s.e.}(p_e) = \frac{1}{(25 + 0)^2}$$
$$\times \sqrt{(25 + 10 + 0)(25 \times 10 + 25 \times 0 + 10 \times 0) - 25 \times 10 \times 0}$$
$$= .15.$$

8.2. MATCHED PAIRS: MORE THAN DICHOTOMOUS OUTCOME

Often the response of a subject to treatment or the degree to which he possesses a factor may be graded more finely than on the simple presence-absence dichotomy considered in the preceding section. Response to treatment, for example, may be graded as improved, essentially unchanged, or worse. Cigarette smoking, as another example, may be graded as none at all, between one and ten cigarettes per day, between 11 and 20 cigarettes per day, and 21 or more cigarettes per day. When the samples being compared are not matched, the methods of Chapter 9 may be applied. Here we consider the case of matched pairs, both members of which are classified into one of $k(>2)$ mutually exclusive categories.

Table 8.5 demonstrates the appropriate presentation of the data. Each entry in the table represents a number of pairs. For example, $n_{..}$ is the total

Table 8.5. Data on k outcomes from matched pairs

Outcome Category for Cases	Outcome Category for Controls				
	1	2	$\cdots$	k	Total
1	n_{11}	n_{12}	$\cdots$	n_{1k}	$n_{1.}$
2	n_{21}	n_{22}	$\cdots$	n_{2k}	$n_{2.}$
.					
.					
.					
k	n_{k1}	n_{k2}	$\cdots$	n_{kk}	$n_{k.}$
Total	$n_{.1}$	$n_{.2}$	$\cdots$	$n_{.k}$	$n_{..}$

number of matched pairs, $n_{1.}$ is the number of pairs in which the case was in category 1, $n_{.2}$ is the number in which the control was in category 2, and n_{12} is the number in which the case was in category 1 and the control in category 2. The differences between the cases and controls are represented by the k differences $d_1 = (n_{1.} - n_{.1})$, $d_2 = (n_{2.} - n_{.2})$, ..., $d_k = (n_{k.} - n_{.k})$. Clearly, these differences do not depend on the quantities $n_{11}, n_{22}, ..., n_{kk}$, the numbers of pairs both of whose members had outcomes in the same category.

Complicated test statistics for assessing the significance of the k differences $d_1, d_2, ..., d_k$ have been proposed by Bhapkar (1966), Grizzle, Starmer and Koch (1969), and Ireland, Ku, and Kullback (1969). A simpler test statistic, but one that still requires the inversion of a matrix, has been proposed by Stuart (1955) and Maxwell (1970). A simple expression for the Stuart-Maxwell statistic when $k = 3$ has been derived by Fleiss and Everitt (1971).

For $k = 3$, define

$$\bar{n}_{ij} = \frac{n_{ij} + n_{ji}}{2}. \tag{8.8}$$

The statistic

$$\chi^2 = \frac{\bar{n}_{23}d_1^2 + \bar{n}_{13}d_2^2 + \bar{n}_{12}d_3^2}{2(\bar{n}_{12}\bar{n}_{13} + \bar{n}_{12}\bar{n}_{23} + \bar{n}_{13}\bar{n}_{23})} \tag{8.9}$$

may be referred to tables of chi square with two degrees of freedom (see Table A.1). If χ^2 is significantly large, the inference would be made that the distribution across the categories for the cases differs from the distribution for the controls.

Consider the data of Table 8.6, in which it is assumed that the treatments were assigned to the members of each matched pair at random. The value of the chi square statistic (8.9) is

$$\chi^2 = \frac{\dfrac{5+5}{2}(40-60)^2 + \dfrac{0+10}{2}(40-30)^2 + \dfrac{5+15}{2}(20-10)^2}{2\left(\dfrac{5+15}{2} \times \dfrac{0+10}{2} + \dfrac{5+15}{2} \times \dfrac{5+5}{2} + \dfrac{0+10}{2} \times \dfrac{5+5}{2}\right)}$$

$$= \frac{3500}{2 \times 125} = 14.00,$$

which, with two degrees of freedom, is significant beyond the .001 level. It may therefore be concluded that the outcome distribution of patients given the new treatment is different from that of patients given the standard treatment.

When, as in this example, a significant difference is found between the two distributions, the next step in the analysis would be to find those single

Table 8.6. Hypothetical data to illustrate the Stuart-Maxwell test

New Treatment	Standard Treatment			
	Improved	No Change	Worse	Total
Improved	35	5	0	40
No change	15	20	5	40
Worse	10	5	5	20
Total	60	30	10	100

categories (in the case of more than three categories, possibly those combinations of categories) for which the differences are significant [see Fleiss and Everitt (1971) for a general discussion]. One need only collapse the original table into a 2 × 2 table and apply McNemar's statistic (8.3). The test for significance, however, must incorporate a control over the fact that the chances of erroneously declaring a difference to be significant increase when a number of tests are applied to the same data. An appropriate control in the case we are considering (see Miller, 1966, Section 6.2) is to refer McNemar's chi square statistic to the critical value of chi square with $k - 1$ degrees of freedom.

We illustrate the search for those categories with a significant difference using the data of Table 8.6. To determine whether the proportions who improved under the two treatments were different, we form the following 2 × 2 table. Sixty percent of the patients given the standard treatment

Table 8.7. Two-by-two table for comparing rates of improvement: data from Table 8.6

New Treatment	Standard Treatment		
	Improved	Not Improved	Total
Improved	35	5	40
Not improved	25	35	60
Total	60	40	100

improved, while only 40% of those given the new treatment did. The value of McNemar's statistic is

$$\chi^2 = \frac{(|5 - 25| - 1)^2}{5 + 25} = \frac{19^2}{30} = 12.03.$$

The critical value of chi square with two degrees of freedom for a significance level of .05 (see Table A.1) is 5.99. Since the obtained value of McNemar's chi square exceeds 5.99, we may infer that the improvement rate under the new treatment is less than that under the standard.

Table 8.8 contrasts the rates of worsening under the two treatments.

Table 8.8. *Two-by-two table for comparing rates of worsening: data from Table 8.6*

	Standard Treatment		
New Treatment	Worse	Not Worse	Total
Worse	5	15	20
Not worse	5	75	80
Total	10	90	100

Ten percent of the patients given the standard treatment, as opposed to 20% of those given the new one, ended up worse than initially. The value of McNemar's chi square statistic is

$$\chi^2 = \frac{(|15 - 5| - 1)^2}{15 + 5} = \frac{9^2}{20} = 4.05.$$

Since this value fails to exceed 5.99, the proper criterion when a 3×3 table has been collapsed into a 2×2, the inference we draw is that the rates of worsening under the two treatments may be equal.

8.3. MULTIPLE CONTROLS: THE CASE OF TWO SAMPLES

Occasionally, two matched samples may be generated by matching each case (or each patient given a new treatment) with more than one control (or with more than one patient given a standard treatment). Matching with multiple controls is especially advantageous when the number of potential control subjects is large relative to the number of available cases, and when little effort needs to be expended in obtaining the necessary information.

We assume that each subject is characterized by either the presence or the absence of some factor or outcome. A general method of analysis, valid even when the number of controls varies from one case to another, was originally derived by Mantel and Haenszel (1959). An alternative but more complex method of analysis in the general case is due to Cox (1966). Here we assume that each case is matched with the same number, say $m - 1$, of controls, and consider only the Mantel-Haenszel method.

Suppose that there are a total of N matched m-tuples, each containing one case and $m - 1$ controls. In the ith m-tuple ($i = 1, \ldots, N$), let x_i denote the number of controls who had the factor (so that x_i may equal $0, 1, \ldots,$ or

$m - 1$), and let n_i denote the total number of subjects—including the case and controls—who had the factor. Thus, if the case in the ith m-tuple had the factor, then $n_i = x_i + 1$; if he did not have the factor, then $n_i = x_i$.

Define

$$A = \sum_{i=1}^{N} x_i, \tag{8.10}$$

the total number of control subjects who had the factor, and define

$$B = \sum_{i=1}^{N} n_i, \tag{8.11}$$

the total number of either kind of subject who had the factor. Note that the total number of cases who had the factor is $B - A$. The rate at which the factor is present among the controls is

$$p_1 = \frac{A}{N(m - 1)}, \tag{8.12}$$

and the rate at which it is present among the cases is

$$p_2 = \frac{B - A}{N}. \tag{8.13}$$

In order to test the significance of the difference between p_1 and p_2, the statistic

$$\chi^2 = \left(\frac{p_2 - p_1}{\text{s.e.}(p_2 - p_1)}\right)^2 = \frac{((m - 1)B - mA)^2}{mB - \sum_{i=1}^{N} n_i^2} \tag{8.14}$$

may be referred to tables of chi square with one degree of freedom (see Miettinen, 1969, and Pike and Morrow, 1970). Miettinen (1969) has studied the power of the test based on (8.14), and has given criteria (in terms of reducing cost) for deciding on an appropriate value for $m - 1$, the number of controls per case.

The data in Table 8.9 are used to illustrate this analysis. Suppose that $N = 10$ matched triples ($m = 3$) were studied, implying $m - 1 = 2$ controls per case. The proportion of controls having the factor (8.12) is

$$p_1 = \frac{7}{10 \times 2} = .35$$

and the proportion of cases having the factor (8.13) is

$$p_2 = \frac{8}{10} = .80.$$

Table 8.9.   Outcome data from matched triples

Triple	Case Has Factor*	Number of Controls with Factor ($= x_i$)	Total Having Factor ($= n_i$)	n_i^2
1	1	2	3	9
2	1	1	2	4
3	1	1	2	4
4	1	1	2	4
5	1	1	2	4
6	1	0	1	1
7	1	0	1	1
8	1	0	1	1
9	0	1	1	1
10	0	0	0	0
Total	8 ($= B - A$)	7 ($= A$)	15 ($= B$)	29

* 1 = yes, 0 = no.

The value of the statistic (8.14) for testing the significance of the difference between these two proportions is

$$\chi^2 = \frac{(2 \times 15 - 3 \times 7)^2}{3 \times 15 - 29} = \frac{9^2}{16} = 5.06.$$

Since this value exceeds 3.84, the value of chi square with one degree of freedom needed for significance at the .05 level, the inference may be drawn that the proportion of cases having the factor is larger than the proportion of controls having it.

The Mantel-Haenszel estimate of the odds ratio (1959, p. 736) is

$$o = \frac{(m - 1)(B - A) - \sum_{i=1}^{N} x_i(n_i - x_i)}{A - \sum_{i=1}^{N} x_i(n_i - x_i)}. \qquad (8.15)$$

The quantity $\sum x_i(n_i - x_i)$ is obtained by restricting attention to those m-tuples in which the case had the factor, and simply adding the numbers of controls in them who had the factor.

For the data of Table 8.9, only the first eight triples were such that the case had the factor. The total number of controls in those eight triples who had the factor is 6 ($= \sum x_i(n_i - x_i)$), so that the estimated odds ratio (8.15) is

$$o = \frac{2 \times 8 - 6}{7 - 6} = 10.0.$$

Miettinen (1970b) presents an alternative method for estimating the odds ratio in the case of matched m-tuples [but one more complicated than (8.15)], and gives approximate expressions for the standard error of the estimate.

8.4. MULTIPLE CONTROLS: THE CASE OF m SAMPLES

In the preceding section we considered the case where the $m - 1$ controls for each case formed a homogeneous group. In this section we consider the comparison of m distinct matched samples, but again restrict attention to the case where only two outcomes are of interest (see Koch and Reinfurt, 1971, for the general case).

The case being considered would arise in a comparative prospective study in which, for example, a number of quadruples ($m = 4$) of subjects would be matched on sex and age, under the restriction that one member did not smoke cigarettes, another smoked between one and ten cigarettes per day, a third smoked between 11 and 20 cigarettes per day, and a fourth smoked 21 or more cigarettes per day. The proportions of subjects from the four resulting matched samples who develop a disease would then be compared using the methods of this section. An example of a retrospective study with three matched samples (lung cancer patients, other patients, and community controls) is one by Doll and Hill (1952).

The methods of this section are also applicable to the results of a controlled trial in which $m > 2$ treatments are compared by grouping together a number of sets of m similar patients each, and randomly assigning the treatments to the patients within each matched m-tuple. The methods are applicable, too, when each of a sample of subjects is studied under m different conditions. An example is the comparison of the proportions positive associated with m diagnostic tests, when each test is applied to each patient in the sample.

Table 8.10 illustrates the presentation of the data resulting from the study of m matched samples, with N observations in each sample. In Table 8.10, each X is either 0 (if the response is negative) or 1 (if the response is positive). Thus, for example, S_1 represents the total number of positives from the first m-tuple, T_1 represents the total number of positives from the first sample, and T represents the overall total number of positives.

Define

$$p_j = \frac{T_j}{N}, \tag{8.16}$$

the proportion of subjects from the jth sample who were positive;

$$P_n = \frac{S_n}{m}, \tag{8.17}$$

Table 8.10.   Presentation of data from m
matched samples

m-tuple	Sample 1	2	$\cdots$	m	Total
1	X_{11}	X_{12}	$\cdots$	X_{1m}	S_1
2	X_{21}	X_{22}	$\cdots$	X_{2m}	S_2
.					
.					
.					
N	X_{N1}	X_{N2}	$\cdots$	X_{Nm}	S_N
Total	T_1	T_2	$\cdots$	T_m	T
Proportion	p_1	p_2	$\cdots$	p_m	$\bar{p}$

the proportion of positives in the nth m-tuple; and

$$\bar{p} = \frac{1}{m} \sum_{j=1}^{m} p_j = \frac{1}{N} \sum_{n=1}^{N} P_n = \frac{T}{Nm}, \tag{8.18}$$

the overall proportion positive. Interest is in whether the proportions $p_1, \ldots, p_m$ differ significantly. The following statistic, due to Cochran (1950), may be used to test for the significance of the differences among the m proportions:

$$Q = \frac{N^2(m-1)}{m} \times \frac{\sum_{j=1}^{m}(p_j - \bar{p})^2}{N\bar{p}(1-\bar{p}) - \sum_{n=1}^{N}(P_n - \bar{p})^2}$$

$$= (m-1) \times \frac{m\sum_{j=1}^{m} T_j^2 - T^2}{mT - \sum_{n=1}^{N} S_n^2}. \tag{8.19}$$

The value of (8.19) may be referred to tables of chi square with $m - 1$ degrees of freedom (see Table A.1).

Consider the data of Table 8.11, originally reported by Fleiss (1965a). The proportions p_j of patients judged to have religious preoccupations vary from a low of 0 to a high of .375. The value of Q (8.19) for testing whether this variation can be attributed to chance or whether it represents real differences among the raters is

$$Q = 7 \times \frac{8(1^2 + 0^2 + \cdots + 0^2 + 3^2) - 15^2}{8 \times 15 - (0^2 + 1^2 + \cdots + 0^2 + 6^2)} = 14.71.$$

Table 8.11. *Judgments by eight raters as to presence or absence* of religious preoccupation in eight patients*

	Rater								Total
Patient	1	2	3	4	5	6	7	8	$(= S_n)$
1	0	0	0	0	0	0	0	0	0
2	0	0	0	0	1	0	0	0	1
3	0	0	0	0	0	0	0	0	0
4	0	0	0	0	0	0	0	0	0
5	0	0	1	0	0	1	0	1	3
6	0	0	1	1	1	1	0	1	5
7	0	0	0	0	0	0	0	0	0
8	1	0	1	1	1	1	0	1	6
Total $(= T_j)$	1	0	3	2	3	3	0	3	$15 (= T)$
Proportion $(= p_j)$	.125	0	.375	.250	.375	.375	0	.375	$.234 (= \bar{p})$

* 1 or 0, respectively.

The validity of calculating Q for many ratings on the same subjects has been established by Fleiss (1965b).

Referring to Table A.1 with $m - 1 = 7$ degrees of freedom, we find that Q must exceed 14.07 in order for the variation to be declared significant at the .05 level. Since our obtained value of 14.71 exceeds the critical value, we infer that the raters differ in their judgments of religious preoccupation.

Having found significant variation, our next step would be to try to identify those samples or groups of samples (in our example, those raters or groups of raters) which differed. A device that is frequently useful is to *partition* Q into separate components, each of which measures a specified source of variability. The general method for partitioning a chi square statistic is described by Maxwell (1961, Chapter 3). Here we illustrate the method for the statistic (8.19).

Suppose that the m samples represent two groups, with m_1 samples in the first group and m_2 in the second. In the first example given at the beginning of the section, one group consists of the single sample of nonsmokers (so that $m_1 = 1$) and the other of the three samples of cigarette smokers (so that $m_2 = 3$). Define

$$U_1 = \sum_{j=1}^{m_1} T_j, \tag{8.20}$$

the total number of positives in the first group of samples, and

$$U_2 = \sum_{j=m_1+1}^{m} T_j, \tag{8.21}$$

the total number of positives in the second group. The statistic for testing whether the proportion positive in the first group differs significantly from that in the second is

$$Q_{\text{diff}} = \frac{(m-1)}{m_1 m_2} \times \frac{(m_2 U_1 - m_1 U_2)^2}{mT - \sum_{n=1}^{N} S_n^2} . \tag{8.22}$$

The statistic Q_{diff} has one degree of freedom.

Consider again the data of Table 8.11. The first five raters were from New York and the last three were from Kentucky. They therefore form two natural groups, one containing $m_1 = 5$ raters and the other $m_2 = 3$. It is reasonable to inquire whether the two groups of raters differ in their judgments. The total number of positive ratings by the raters in the first group is

$$U_1 = 1 + 0 + 3 + 2 + 3 = 9,$$

and the total number of positive ratings by those in the second group is

$$U_2 = 3 + 0 + 3 = 6.$$

The value of Q_{diff} (8.22) is

$$Q_{\text{diff}} = \frac{7}{5 \times 3} \times \frac{(3 \times 9 - 5 \times 6)^2}{120 - 71} = 0.09,$$

indicating a negligible difference between the New York raters as a group and the Kentucky raters as a group.

The next step in the analysis would be to compare the m_1 samples within group 1 by means of the statistic

$$Q_{\text{group 1}} = \frac{m(m-1)}{m_1} \times \frac{m_1 \sum_{j=1}^{m_1} T_j^2 - U_1^2}{mT - \sum_{n=1}^{N} S_n^2} , \tag{8.23}$$

and to compare the m_2 samples within group 2 by means of the statistic

$$Q_{\text{group 2}} = \frac{m(m-1)}{m_2} \times \frac{m_2 \sum_{j=m_1+1}^{m} T_j^2 - U_2^2}{mT - \sum_{n=1}^{N} S_n^2} . \tag{8.24}$$

The statistic $Q_{\text{group 1}}$ has $m_1 - 1$ degrees of freedom and the statistic $Q_{\text{group 2}}$ has $m_2 - 1$ degrees of freedom. It may be checked that

$$Q = Q_{\text{diff}} + Q_{\text{group 1}} + Q_{\text{group 2}}.$$

In addition, note that the three degrees of freedom for the Q statistics of

(8.22) to (8.24), namely 1, $m_1 - 1$, and $m_2 - 1$, sum to $m_1 + m_2 - 1 = m - 1$, the degrees of freedom in the overall Q (8.19).

For the data of Table 8.11, the differences among the five New York raters are assessed by

$$Q_{\text{group 1}} = \frac{8 \times 7}{5} \times \frac{5(1^2 + 0^2 + 3^2 + 2^2 + 3^2) - 9^2}{120 - 71}$$

$$= 7.77,$$

which with $5 - 1 = 4$ degrees of freedom fails to reach significance at the .05 level. The differences among the three Kentucky raters are assessed by

$$Q_{\text{group 2}} = \frac{8 \times 7}{3} \times \frac{3(3^2 + 0^2 + 3^2) - 6^2}{120 - 71}$$

$$= 6.86,$$

which with $3 - 1 = 2$ degrees of freedom is significant at the .05 level. Note that

$$Q_{\text{diff}} + Q_{\text{group 1}} + Q_{\text{group 2}} = 0.09 + 7.77 + 6.86 = 14.72,$$

equal except for rounding errors to the overall value of Q, 14.71.

Cochran (1950, p. 265) suggests a slightly different approach to the partitioning of Q. Its effect is to reduce slightly the magnitudes of $Q_{\text{group 1}}$ and of $Q_{\text{group 2}}$ (see also Tate and Brown, 1970). The conclusions for the data of Table 8.11 are the same for both methods of partitioning: there are differences among the judgments of the eight raters, arising essentially from the variability among the three Kentucky raters.

The Q statistic (8.19), like the test statistics presented in the three preceding sections, is unaffected by the deletion of those m-tuples in which either all m responses were positive or all m were negative. Seeger and Gabrielsson (1968) and Tate and Brown (1970) have studied the accuracy of the chi square approximation to the distribution of Q. It seems that the approximation is adequate provided the product of the number of samples ($= m$) and the number of m-tuples remaining after deletion of those in which all responses were the same is at least 24. For the data of Table 8.11, four patients (numbers 1, 3, 4, and 7) were such that the ratings were identical. The product of $m = 8$ and the remaining number of patients ($= 4$) is 32, indicating that the approximation was adequate.

A quite different approach from Cochran's (1950) to the comparison of m matched samples is due to Bennett (1967, 1968). The reader is referred to his two papers for the test statistics he derived (more complicated than the Q statistic) and for their powers.

8.5. ADVANTAGES AND DISADVANTAGES OF MATCHING

In comparative prospective and retrospective studies, the matching of subjects is often necessary to assure that the samples being contrasted are similar with respect to characteristics associated with the factors being studied (see, e.g., Billewicz, 1965, and Miettinen, 1970a). The possible gain in efficiency due to the study of matched samples (i.e., increase in the power of the test of significance and increase in the precision of the estimated degree of association) therefore assumes lesser importance, but some results are available.

Cochran (1950) and Worcester (1964) have shown that matching is not guaranteed to increase efficiency. It can be expected to do so only when the characteristics being matched are strongly associated with the factors under study. When the characteristics used for matching are only slightly, or are not at all associated with the factors under study, Youkeles (1963) has shown that efficiency may even be lost. When the sample sizes exceed 30, however, matching on irrelevant characteristics seems not to affect efficiency.

In the context of controlled comparative trials with random assignment of treatments to subjects, on the other hand, the purpose of matching is mainly the increase of efficiency. Chase (1968) has shown that matching is at least as efficient as no matching except for small sample sizes. Billewicz (1964, 1965), however, has indicated that the increase in efficiency is frequently only limited, and may not be worth the effort involved in securing adequate matches.

He has shown how the length of time required to complete a study increases either as the number of matching characteristics increases or as the relative frequencies of some categories of the matching characteristics decrease. With too many matching characteristics, or with only a few but some containing too many categories, the investigator may find himself with a large proportion of subjects left unmatched at the end of the study.

In view of the need to assure comparability in comparative prospective and retrospective studies, matching in these contexts is often called for. Matching should, however, be on a small number of characteristics (rarely more than four and preferably no more than two), with each defined by a small number of categories (with respect to age, e.g., matching by 10-year intervals should frequently suffice).

In view of the limited gain in efficiency due to matching in controlled comparative trials, however, and in view of the likelihood that the time required to complete such a trial may increase inordinately, matching should be undertaken only on characteristics strongly associated with the response being studied, and then on no more than one or two of them, and then again with each subdivided into only a small number of categories.

Problem 8.1. Two diagnosticians both diagnosed each of a sample of 100 patients:

	Diagnostician A			
Diagnostician B	Schizophrenia	Affective	Other	Total
Schizophrenia	40	6	4	50
Affective	20	6	4	30
Other	10	8	2	20
Total	70	20	10	100

(*a*) Use the statistic given in (8.9) to test whether the diagnostic distribution of diagnostician A is the same as that of diagnostician B.

(*b*) Do the two diagnosticians differ significantly in the proportions of patients they diagnose schizophrenic? affectively ill? other?

Problem 8.2. When not corrected for continuity, McNemar's chi square statistic is given by

$$\chi_u^2 = \frac{(b - c)^2}{b + c}.$$

Prove that, when $m = 2$, the expression for χ^2 given by (8.14) is equal to that of χ_u^2. (*Hint.* Refer to Table 8.1 for notation. Prove that, when $m = 2$, A (8.10) equals $a + c$, B (8.11) equals $2a + b + c$, and $\sum n_i^2$ equals $4a + b + c$.)

Problem 8.3. Prove that, when $m = 2$, the value of Q given by (8.19) is equal to that of χ_u^2. (*Hint.* Prove that, when $m = 2$, $T_1 = a + c$, $T_2 = a + b$, $T = 2a + b + c$, and $\sum S_n^2 = 4a + b + c$.)

REFERENCES

Bennett, B. M. (1967). Tests of hypotheses concerning matched samples. *J. roy. statist. Soc., Ser. B*, **29**, 468–474.

Bennett, B. M. (1968). Note on X^2 tests for matched samples. *J. roy. statist. Soc., Ser. B*, **30**, 368–370.

Bennett, B. M. and Underwood, R. E. (1970). On McNemar's test for the 2 × 2 table and its power function. *Biometrics*, **26**, 339–343.

Bhapkar, V. P. (1966). A note on the equivalence of two test criteria for hypotheses in categorical data. *J. Amer. statist. Assoc.*, **61**, 228–235.

Billewicz, W. Z. (1964). Matched samples in medical investigations. *Brit. J. prev. soc. Med.*, **18**, 167–173.

Billewicz, W. Z. (1965). The efficiency of matched samples: An empirical investigation. *Biometrics*, **21**, 623–644.

Bross, I. D. J. (1969). How case-for-case matching can improve design efficiency. *Amer. J. Epidemiol.*, **89**, 359–363.

Chase, G. R. (1968). On the efficiency of matched pairs in Bernoulli trials. *Biometrika*, **55**, 365–369.

Cochran, W. G. (1950). The comparison of percentages in matched samples. *Biometrika*, **37**, 256–266.

Cornfield, J. and Haenszel, W. (1960). Some aspects of retrospective studies. *J. chronic Dis.*, **11**, 523–534.

Cox, D. R. (1966). A simple example of a comparison involving quantal data. *Biometrika*, **53**, 215–220.

Doll, R. and Hill, A. B. (1952). A study of the etiology of carcinoma of the lung. *Brit. med. J.*, **2**, 1271–1286.

Edwards, A. L. (1948). Note on the "correction for continuity" in testing the significance of the difference between correlated proportions. *Psychometrika*, **13**, 185–187.

Fleiss, J. L. (1965a). Estimating the accuracy of dichotomous judgments. *Psychometrika*, **30**, 469–479.

Fleiss, J. L. (1965b). A note on Cochran's Q test. *Biometrics*, **21**, 1008–1010.

Fleiss, J. L. and Everitt, B. S. (1971). Comparing the marginal totals of square contingency tables. *Brit. J. math. statist. Psychol.*, **24**, 117–123.

Grizzle, J. E., Starmer, C. F. and Koch, G. G. (1969). Analysis of categorical data by linear models. *Biometrics*, **25**, 489–504.

Hill, A. B. (1962). *Statistical methods in clinical and preventive medicine*. New York: Oxford University Press.

Ireland, C. T., Ku, H. H. and Kullback, S. (1969). Symmetry and marginal homogeneity of an $r \times r$ contingency table. *J. Amer. statist. Assoc.*, **64**, 1323–1341.

Koch, G. G. and Reinfurt, D. W. (1971). The analysis of categorical data from mixed models. *Biometrics*, **27**, 157–173.

McNemar, Q. (1947). Note on the sampling error of the difference between correlated proportions or percentages. *Psychometrika*, **12**, 153–157.

Mantel, N. and Haenszel, W. (1959). Statistical aspects of the analysis of data from retrospective studies of disease. *J. natl. Cancer Inst.*, **22**, 719–748.

Maxwell, A. E. (1961). *Analysing qualitative data*. London: Methuen.

Maxwell, A. E. (1970). Comparing the classification of subjects by two independent judges. *Brit. J. Psychiatr.*, **116**, 651–655.

Miettinen, O. S. (1968). The matched pairs design in the case of all-or-none responses. *Biometrics*, **24**, 339–352.

Miettinen, O. S. (1969). Individual matching with multiple controls in the case of all-or-none responses. *Biometrics*, **25**, 339–355.

Miettinen, O. S. (1970a). Matching and design efficiency in retrospective studies. *Amer. J. Epidemiol.*, **91**, 111–118.

Miettinen, O. S. (1970b). Estimation of relative risk from individually matched series. *Biometrics*, **26**, 75–86.

Miller, R. G. (1966). *Simultaneous statistical inference*. New York: McGraw-Hill.

Mosteller, F. (1952). Some statistical problems in measuring the subjective response to drugs. *Biometrics*, **8**, 220–226.

Pike, M. C. and Morrow, R. H. (1970). Statistical analysis of patient-control studies in epidemiology: Factor under investigation an all-or-none variable. *Brit. J. prev. soc. Med.*, **24**, 42–44.

Seeger, P. and Gabrielsson, A. (1968). Applicability of the Cochran Q test and the F test for statistical analysis of dichotomous data for dependent samples. *Psychol. Bull.*, **69**, 269–277.

Stuart, A. (1955). A test for homogeneity of the marginal distribution in a two-way classification. *Biometrika*, **42**, 412–416.

Stuart, A. (1957). The comparison of frequencies in matched samples. *Brit. J. statist. Psychol.*, **10**, 29–32.

Tate, M. W. and Brown, S. M. (1970). Note on the Cochran Q test. *J. Amer. statist. Assoc.*, **65**, 155–160.

Worcester, J. (1964). Matched samples in epidemiologic studies. *Biometrics*, **20**, 840–848.

Youkeles, L. H. (1963). Loss of power through ineffective pairing of observations in small two-treatment all-or-none experiments. *Biometrics*, **19**, 175–180.

CHAPTER 9

The Comparison of Proportions from Many Samples

With only a few exceptions, we have restricted our attention to the comparison of two proportions. In this chapter we consider the comparison of a number of proportions. In Section 9.1 we study the analysis of an $m \times 2$ contingency table, where $m > 2$ and where there is no necessary ordering to the m categories or groups. Sections 9.2 to 9.4 are devoted to the case where an intrinsic ordering to the m categories exists. We consider in Section 9.2 the hypothesis that the proportions vary monotonically (i.e., steadily increase or steadily decrease) with m quantitatively ordered categories, and in Section 9.3 that they vary monotonically with m qualitatively ordered categories. In Section 9.4 we continue to consider the case of a qualitative ordering to the categories, but do not restrict the proportions to vary monotonically.

The procedures of this chapter are appropriate to each of the three methods of sampling presented previously (see Section 2.1). In method III sampling, the m samples represent groups treated by m different treatments, with subjects assigned to groups at random. In method II sampling, the investigator selects either prespecified numbers of subjects from each of the m groups or prespecified numbers with and without the outcome characteristic. In method I sampling, these numbers become known only after the study is completed. As was the case for the comparison of $m = 2$ samples (see Sections 6.1 and 6.2), method II sampling with equal sample sizes is superior in terms of power and precision to method I sampling when $m > 2$.

9.1. THE COMPARISON OF m PROPORTIONS

Suppose that m independent samples of subjects are studied, with each subject characterized by the presence or absence of some characteristic. The

Table 9.1. Proportions from m independent samples

Sample	Total in Sample	Number with Characteristic	Number without Characteristic	Proportion with Characteristic
1	$n_{1.}$	n_{11}	n_{12}	p_1
2	$n_{2.}$	n_{21}	n_{22}	p_2
.				
.				
.				
m	$n_{m.}$	n_{m1}	n_{m2}	p_m
Overall	$n_{..}$	$n_{.1}$	$n_{.2}$	$\bar{p}$

resulting data might be presented as in Table 9.1. In Table 9.1,

$$p_i = \frac{n_{i1}}{n_{i.}} \tag{9.1}$$

and

$$\bar{p} = \frac{n_{.1}}{n_{..}} = \frac{\sum n_{i.} p_i}{\sum n_{i.}}. \tag{9.2}$$

For testing the significance of the differences among the m proportions, the value of

$$\chi^2 = \sum_{i=1}^{m} \sum_{j=1}^{2} \frac{(n_{ij} - n_{i.} n_{.j}/n_{..})^2}{n_{i.} n_{.j}/n_{..}} \tag{9.3}$$

may be referred to tables of chi square (see Table A.1) with $m - 1$ degrees of freedom. An equivalent and more suggestive formula for the test statistic is

$$\chi^2 = \frac{1}{\bar{p}\bar{q}} \sum_{i=1}^{m} n_{i.}(p_i - \bar{p})^2, \tag{9.4}$$

where $\bar{q} = 1 - \bar{p}$.

Consider, as an example, the following data from four studies cited by Dorn (1954). In each study, the number of smokers among lung cancer

Table 9.2. Smoking status among lung cancer patients in four studies

Study	Number of Patients	Number of Smokers	Proportion of Smokers
1	86 $(= n_{1.})$	83	.965 $(= p_1)$
2	93 $(= n_{2.})$	90	.968 $(= p_2)$
3	136 $(= n_{3.})$	129	.949 $(= p_3)$
4	82 $(= n_{4.})$	70	.854 $(= p_4)$
Overall	397 $(= n_{..})$	372	.937 $(= \bar{p})$

patients was recorded. For these data, the value of χ^2 (9.4) is

$$\chi^2 = \frac{1}{.937 \times .063} (86 \times (.965 - .937)^2 + 93 \times (.968 - .937)^2$$
$$+ 136 \times (.949 - .937)^2 + 82 \times (.854 - .937)^2)$$
$$= 12.56 \tag{9.5}$$

which, with three degrees of freedom, is significant at the .01 level.

Having found the proportions to differ significantly, one would next proceed to identify the samples or groups of samples that contributed to the significant difference. Methods for isolating sources of significant differences in the context of a general contingency table are given by Irwin (1949), Lancaster (1950), Kimball (1954), Kastenbaum (1960), and Castellan (1965). Here we illustrate the method for the $m \times 2$ table.

Suppose that the m samples are partitioned into two groups, the first containing m_1 samples and the second m_2, where $m_1 + m_2 = m$. Define

$$n_{(1)} = \sum_{i=1}^{m_1} n_i. \tag{9.6}$$

to be the total number of subjects in the first group of samples and

$$n_{(2)} = \sum_{i=m_1+1}^{m} n_i. \tag{9.7}$$

to be the total number of subjects in the second group.

Let the proportion in the first group be denoted $\bar{p}_1$, where

$$\bar{p}_1 = \frac{\sum_{i=1}^{m_1} n_i.p_i}{n_{(1)}}, \tag{9.8}$$

and that in the second group be denoted $\bar{p}_2$, where

$$\bar{p}_2 = \frac{\sum_{i=m_1+1}^{m} n_i.p_i}{n_{(2)}}. \tag{9.9}$$

Then

$$\chi^2_{\text{diff}} = \frac{1}{\bar{p}\bar{q}} \times \frac{n_{(1)}n_{(2)}}{n_{..}} (\bar{p}_1 - \bar{p}_2)^2, \tag{9.10}$$

with one degree of freedom, may be used to test for the significance of the difference between $\bar{p}_1$ and $\bar{p}_2$. The statistic

$$\chi^2_{\text{group 1}} = \frac{1}{\bar{p}\bar{q}} \sum_{i=1}^{m_1} n_i.(p_i - \bar{p}_1)^2, \tag{9.11}$$

with $m_1 - 1$ degrees of freedom, may be used to test the significance of the differences among the m_1 proportions in the first group, and the statistic

$$\chi^2_{\text{group 2}} = \frac{1}{\bar{p}\bar{q}} \sum_{i=m_1+1}^{m} n_{i.}(p_i - \bar{p}_2)^2, \qquad (9.12)$$

with $m_2 - 1$ degrees of freedom, may be used to test the significance of the differences among the m_2 proportions in the second group. It may be checked that the three statistics given by (9.10) to (9.12) sum to the overall value of χ^2 (9.4).

If $\bar{p}_1$ and $\bar{p}_2$ differ appreciably, then the product $\bar{p}_1\bar{q}_1 = \bar{p}_1(1 - \bar{p}_1)$ should replace $\bar{p}\bar{q}$ in (9.11), and $\bar{p}_2\bar{q}_2 = \bar{p}_2(1 - \bar{p}_2)$ should replace $\bar{p}\bar{q}$ in (9.12). These adjustments usually have little effect on the magnitudes of χ^2.

A more serious modification is called for, however, if the partitioning of the samples into groups is suggested by the data instead of being planned beforehand. Of the four samples in Table 9.2, for example, the first three appear, on the basis of the similarity of their proportions, to form one homogeneous group, whereas the fourth sample seems to stand by itself as a second group. To control for the erroneous inferences possible by making comparisons suggested by the data, each of the χ^2 values in (9.10) to (9.12) should be referred to the critical value of chi square with $m - 1$ degrees of freedom and *not* to the critical values of chi square with 1, $m_1 - 1$, and $m_2 - 1$ degrees of freedom (Miller, 1966, Section 6.2).

For the data of Table 9.2, for example, the first group of $m_1 = 3$ studies consists of

$$n_{(1)} = 86 + 93 + 136 = 315$$

lung cancer patients, of whom the proportion smoking is

$$\bar{p}_1 = \frac{83 + 90 + 129}{315} = .959.$$

The second group of $m_2 = 1$ study alone consists of $n_{(2)} = 82$ patients, of whom the proportion smoking is $\bar{p}_2 = .854$.

The significance of the difference between $\bar{p}_1$ and $\bar{p}_2$ is assessed by the magnitude of χ^2_{diff} (9.10):

$$\chi^2_{\text{diff}} = \frac{1}{.937 \times .063} \times \frac{315 \times 82}{397} (.959 - .854)^2$$

$$= 12.15. \qquad (9.13)$$

The significance of the differences among p_1, p_2, and p_3—all from group 1—

is assessed by the magnitude of $\chi^2_{\text{group 1}}$ (9.11):

$$\chi^2_{\text{group 1}} = \frac{1}{.937 \times .063} (86 \times (.965 - .959)^2$$
$$+ 93 \times (.968 - .959)^2 + 136 \times (.949 - .959)^2)$$
$$= 0.41. \tag{9.14}$$

Because group 2 consists of but a single study sample, the statistic $\chi^2_{\text{group 2}}$ (9.12) is inapplicable here.

Note first of all that

$$\chi^2_{\text{diff}} + \chi^2_{\text{group 1}} = 12.15 + 0.41 = 12.56,$$

equal to the value of the overall chi square statistic given in (9.5). Note next that, with $\bar{p}_1 \bar{q}_1 = .959 \times .041$ replacing $.937 \times .063$ in (9.14), the value of $\chi^2_{\text{group 1}}$ increases only slightly, to 0.62. Recall, finally, that the partitioning was suggested by the data and not planned *a priori*. The values of both χ^2_{diff} and $\chi^2_{\text{group 1}}$ must therefore be referred to the critical value of chi square with $m - 1 = 4 - 1 = 3$ degrees of freedom. Since the critical value for a significance level of .05 is 7.81, the conclusion would be that the proportion of smokers among the patients in study 4 differed from the proportions in studies 1 to 3 (because $\chi^2_{\text{diff}} = 12.15 > 7.81$), but that there were no differences among the proportions in studies 1 to 3 (because $\chi^2_{\text{group 1}} = 0.41 < 7.81$).

9.2. GRADIENT IN PROPORTIONS: SAMPLES QUANTITATIVELY ORDERED

The analysis of the preceding section is of quite general validity, but lacks sensitivity when the m samples possess an intrinsic ordering. We assume in this section that the ordering is quantitative; specifically, that a measurement x_i is naturally associated with the ith sample. Data from the National Center for Health Statistics (1970, Tables 1 and 6) are used for illustration (Table 9.3).

Table 9.3. Prevalence of reported insomnia among adult women by age

Age Interval	Number in Interval (= n_i.)	Proportion Reporting Insomnia (= p_i)	Midpoint Age (= x_i)
18–24	534	.280	21.5
25–34	746	.335	30.0
35–44	784	.337	40.0
45–54	705	.428	50.0
55–64	443	.538	60.0
65–74	299	.590	70.0
Overall	3511 (= $n_{..}$)	.393 (= $\bar{p}$)	42.15 (= $\bar{x}$)

Different methods of analysis are called for depending on how the proportions are hypothesized to vary with x (Yates, 1948). Here we consider only the simplest kind of variation, a linear one. Let P_i denote the proportion in the population from which the ith sample was drawn. We hypothesize that

$$P_i = \alpha + \beta x_i, \tag{9.15}$$

where β, the slope of the line, indicates the amount of change in the proportion per unit change in x and α, the intercept, indicates the proportion expected when $x = 0$.

The two parameters of (9.15) may be estimated as follows. Define

$$\bar{x} = \sum_{i=1}^{m} n_i x_i / n_{..}, \tag{9.16}$$

the mean value of x in the given series of data. The slope is estimated as

$$b = \frac{\sum_{i=1}^{m} n_i (p_i - \bar{p})(x_i - \bar{x})}{\sum_{i=1}^{m} n_i (x_i - \bar{x})^2} \tag{9.17}$$

and the intercept as

$$a = \bar{p} - b\bar{x}. \tag{9.18}$$

The calculation of b is simplified somewhat by noting that its numerator is

$$\text{numerator}(b) = \sum_{i=1}^{m} n_i p_i x_i - n_{..} \bar{p}\bar{x} \tag{9.19}$$

and that its denominator is

$$\text{denominator}(b) = \sum_{i=1}^{m} n_i x_i^2 - n_{..} \bar{x}^2. \tag{9.20}$$

A simple expression for the fitted line is

$$\hat{p}_i = \bar{p} + b(x_i - \bar{x}). \tag{9.21}$$

For the data of Table 9.3, $\bar{p} = .393$, $\bar{x} = 42.15$, and

$$b = .0064. \tag{9.22}$$

The fitted straight line becomes

$$\hat{p}_i = .393 + .0064(x_i - 42.15), \tag{9.23}$$

implying an increase of .64 % in the proportion reporting insomnia per yearly increase in age.

It is useful to calculate the estimated proportion corresponding to each x_i in order to compare it with the actual proportion, p_i. If p_i and $\hat{p}_i$ are close in magnitude for all or most categories, then one can conclude that (9.21) provides a good fit to the data, that is, p_i tends to vary linearly with x_i. If p_i and $\hat{p}_i$ tend to differ, then the conclusion is that the association between p_i and x_i is more complicated than a linear one. Having the differences $p_i - \hat{p}_i$ available serves also to identify those categories for which the departures from linearity are greatest.

Table 9.4 contrasts the actual proportions of Table 9.3 with those yielded by (9.23). The fit appears to be a good one.

Table 9.4. Observed and linearly predicted age-specific rates of insomnia

x_i	$n_i.$	p_i	$\hat{p}_i$
21.5	534	.280	.261
30.0	746	.335	.315
40.0	784	.337	.379
50.0	705	.428	.443
60.0	443	.538	.507
70.0	299	.590	.571

A chi square statistic due to Cochran (1954) and Armitage (1955) is available for testing whether the association between p_i and x_i is a linear one. This chi square statistic is

$$\chi^2_{\text{linearity}} = \sum_{i=1}^{m} n_i.(p_i - \hat{p}_i)^2/\bar{p}\bar{q}. \qquad (9.24)$$

$\chi^2_{\text{linearity}}$ has $m - 2$ degrees of freedom, and the hypothesis of linearity would be rejected if $\chi^2_{\text{linearity}}$ were found to be large. The power of this test was studied by Chapman and Nam (1968).

The calculation of $\chi^2_{\text{linearity}}$ is simplified if one first calculates the statistic

$$\chi^2_{\text{slope}} = b^2 \sum_{i=1}^{m} n_i.(x_i - \bar{x})^2/\bar{p}\bar{q} \qquad (9.25)$$

because it may be shown that

$$\chi^2_{\text{linearity}} = \chi^2 - \chi^2_{\text{slope}}, \qquad (9.26)$$

where χ^2 is given by (9.4). The statistic χ^2_{slope} has one degree of freedom and may be used to test the significance of the slope, b. If χ^2_{slope} is large, the inference is that the slope is significantly different from zero, indicating that there is a tendency for increasing values of x_i to be associated with increasing values of p_i if b is positive, or with decreasing values of p_i if b is negative.

For the data of Table 9.3, the value of the overall chi square statistic (9.4) for testing the hypothesis that the proportion reporting insomnia is constant for all age groups is

$$\chi^2 = 140.72. \tag{9.27}$$

The magnitude of this chi square, which has five degrees of freedom, indicates highly significant differences among the age-specific proportions, but fails to describe the steady increase with age of the proportion reporting insomnia.

The chi square statistic for linearity (9.24) is, for the data from Table 9.4,

$$\chi^2_{\text{linearity}} = \frac{534 \times (.280 - .261)^2 + \cdots + 299 \times (.590 - .571)^2}{.393 \times .607}$$

$$= 10.76 \tag{9.28}$$

which, with four degrees of freedom, is significant at the .05 level. The association with age of the proportion of women reporting insomnia is thus not precisely a linear one, but the departures from linearity (i.e., the differences between the observed and linearly predicted proportions) are sufficiently small to make the hypothesis of linearity reasonable.

The chi square statistic of (9.25) assumes the value

$$\chi^2_{\text{slope}} = \frac{.0064^2 \times 757,319.9025}{.393 \times .607}$$

$$= 130.03, \tag{9.29}$$

which, with one degree of freedom, indicates that the slope of the fitted line, $b = .0064$, is significantly different from zero. The difference between the overall chi square of 140.72 and the chi square for testing the significance of the slope, 130.03, should, by (9.26), be equal to the chi square for linearity, 10.76. Except for errors due to rounding, this is seen to be the case.

The inferences to be drawn from this more detailed chi square analysis are that there is a significant tendency for the proportion of women reporting insomnia to increase steadily with age, and that this tendency is, effectively, a linear one. Had the chi square for linearity been significant at, say, the .01 or .005 level instead of merely at the .05 level, the latter inference would not have been warranted.

A slightly different version of the test statistic (9.25) has been derived by Mantel (1963), who also considers testing for linearity in a number of $m \times 2$ tables.

9.3. GRADIENT IN PROPORTIONS: SAMPLES QUALITATIVELY ORDERED

We assumed in Section 9.2 that the m samples could be ordered on a quantitative scale. We assume in this section and the next that the ordering

Table 9.5. *Hypothetical one-month release rates as a function of initial severity*

Initial Severity	Total	Number Released within One Month	Proportion Released within One Month
Mild	$30 (= n_1)$	25	$.83 (= p_1)$
Moderate	$25 (= n_2)$	22	$.88 (= p_2)$
Serious	$20 (= n_3)$	12	$.60 (= p_3)$
Extreme	$25 (= n_4)$	6	$.24 (= p_4)$
Overall	$100 (= n_{..})$	65	$.65 (= \bar{p})$

is merely qualitative. Suppose, for example, that one has data as in Table 9.5. The value of χ^2 (9.4) for these data is

$$\chi^2 = 28.74 \tag{9.30}$$

with three degrees of freedom, clearly significant beyond the .001 level (see Table A.1).

The inference that the four release rates differ significantly is a valid one, but is clearly insufficient in that it fails to describe the almost steady decline in release rates as initial severity worsens. Because it would have been reasonable to hypothesize beforehand a gradient of release rate with severity, an alternative method of analysis is called for. The method of the preceding section is not appropriate because no numerical values can naturally be assigned to the four levels of severity.

Chassan (1960, 1962) proposed a simple test of the hypothesis that m proportions were arrayed in a prespecified order, but his test was shown by Bartholomew (1963) to lack adequate power. Specifically, Chassan's test may be applied only when the sample proportions are arrayed *without exception* in the same order as hypothesized. It would therefore be inapplicable whenever there were slight departures (as, e.g., for p_1 and p_2 in Table 9.5) from the hypothesized order. A more powerful procedure due to Bartholomew (1959a, 1959b) will be described.

Suppose the hypothesis predicts that the ordering $p_1 > p_2 > \cdots > p_m$ should obtain, but that departures from this ordering are observed. For the proportions in Table 9.5, for example, the ordering $p_1 > p_2 > p_3 > p_4$ was predicted, but instead we obtained $p_1 < p_2$ and then, as predicted, $p_2 > p_3 > p_4$.

When departures are found, weighted averages of those adjacent proportions that are out of order are taken until, when the averages replace the original proportions, the hypothesized ordering is observed. The revised proportions are denoted p'. For the proportions in Table 9.5, the weighted average of p_1 and p_2 must be taken. It is

$$\bar{p}_{1.2} = \frac{30 \times .83 + 25 \times .88}{55} = .85. \qquad (9.31)$$

When p_1 and p_2 are replaced by $\bar{p}_{1.2}$, Table 9.6 results.

Table 9.6. Proportions from Table 9.5 revised to be in hypothesized order

Initial Severity	Total	Revised Proportion
Mild	30 $(= n_1)$	.85 $(= p_1')$
Moderate	25 $(= n_2)$	.85 $(= p_2')$
Serious	20 $(= n_3)$	.60 $(= p_3 = p_3')$
Extreme	25 $(= n_4)$	.24 $(= p_4 = p_4')$
Overall	100 $(= n_{..})$	.65 $(= \bar{p})$

The revised proportions are no longer out of order. If they were, the process would have to be continued. When the process has been completed, the statistic

$$\bar{\chi}^2 = \frac{1}{\bar{p}\bar{q}} \sum_{i=1}^{m} n_{i.}(p_i' - \bar{p})^2 \qquad (9.32)$$

is calculated. For the revised proportions of Table 9.6,

$$\bar{\chi}^2 = \frac{1}{.65 \times .35} (30 \times (.85 - .65)^2 + 25 \times (.85 - .65)^2$$
$$+ 20 \times (.60 - .65)^2 + 25 \times (.24 - .65)^2) = 28.27. \qquad (9.33)$$

The value of $\bar{\chi}^2$ may no longer be referred to tables of chi square, however. Instead, Tables A.6 to A.8 are to be used. When $m = 3$ proportions are compared, calculate

$$c = \sqrt{\frac{n_{1.}n_{3.}}{(n_{1.} + n_{2.})(n_{2.} + n_{3.})}} \qquad (9.34)$$

and enter Table A.6 under the desired significance level, interpolating if necessary. When $m = 4$, calculate

$$c_1 = \sqrt{\frac{n_{1.}n_{3.}}{(n_{1.} + n_{2.})(n_{2.} + n_{3.})}} \qquad (9.35)$$

and

$$c_2 = \sqrt{\frac{n_{2.}n_{4.}}{(n_{2.} + n_{3.})(n_{3.} + n_{4.})}}, \qquad (9.36)$$

and enter Table A.7 under the desired significance level, interpolating in both c_1 and c_2 if necessary. If all sample sizes are equal, and if $m \leq 12$, Table A.8 may be used.

For the data of Table 9.6, for which $m = 4$,

$$c_1 = \sqrt{\frac{30 \times 20}{(30 + 25)(25 + 20)}} = .49$$

and

$$c_2 = \sqrt{\frac{25 \times 25}{(25 + 20)(20 + 25)}} = .56.$$

Visual interpolation in Table A.7 (c_1 is approximately equal to .5 and c_2 is nearly midway between .5 and .6) shows that $\bar{\chi}^2$ would have to exceed 9.0 in order for significance to be declared at the .005 level. The obtained value of $\bar{\chi}^2 = 28.27$ from (9.33) is far beyond this critical value.

What is noteworthy, however, is the comparison of the value just found from Table A.7 with the corresponding value from Table A.1 for the standard chi square test with $m - 1 = 3$ degrees of freedom. If no ordering is hypothesized, χ^2 would have to exceed 12.8 (instead of 9.0) for significance to be declared at the .005 level. Thus, *if the hypothesized ordering actually obtains in the population*, Bartholomew's test is more powerful than the standard chi square test. If the hypothesized ordering is not true, however, the averaging process necessary before the calculation of $\bar{\chi}^2$ (9.32) could well reduce its magnitude to insignificance. Further analyses and generalizations of Bartholomew's test have been made by Barlow, Bartholomew, Bremner, and Brunk (1972).

9.4. SAMPLES QUALITATIVELY ORDERED: RIDIT ANALYSIS

Suppose as in the preceding section that one has available data from a number of samples, with the subjects from each sample distributed across a number of ordered categories. For specificity, let us consider automobile accidents, with the phenomenon studied being the degree of injury sustained by the driver. The degree of injury might be graded from none through severe to fatal. Such a grading is clearly subjective and probably not too reliable. It nevertheless seems preferable to the adoption of the simple dichotomy, little or no injury versus severe or fatal injury, because it both possesses some degree of reliability and succeeds in describing the phenomenon more completely than the cruder yes-no system.

There exists the problem, however, of summarizing the data and making comparisons among different samples in an intelligible way. The standard chi square analysis of Section 9.1 may be performed, but crucial information on the natural ordering of the categories would be lost. The analysis of Section 9.3 would take advantage of this natural ordering, but the required hypothesis of a gradient in the proportions may be too strong for unreliable categories.

A frequently employed device is to number the categories from 0 for the least serious to some highest number for the most serious, and then apply the special chi square analysis of Section 9.2, or even calculate means and standard deviations and then apply t tests or analyses of variance. This device of concocting a seemingly numerical measurement system has many drawbacks. For one thing, one is giving the impression of greater accuracy than really exists. For another, the results one gets depend on the particular system of numbers employed. The choice of a system is by no means a simple one.

Consider again the study of automobile accidents, and suppose that we have seven categories of injury, the first two being None and Mild, and the last two, Critical and Fatal. The straightforward system of numbering assigns the seven integers from 0 to 6 successively to the seven categories. This system is hard to justify, for it implies that the difference between no injury and a mild one is equivalent to the difference between a critical injury and a fatal one. The latter difference is obviously more important, but this greater importance can be picked up only by assigning a value in excess of 6 to the final category. Just what this value should be can, however, only be decided arbitrarily. If an underlying logistic model (see Section 5.4) may be assumed, a procedure due to Snell (1964) is appropriate.

Let us abandon the attempt to quantify the categories, and instead agree to work only with the natural ordering that exists. A technique that takes advantage of this natural ordering, but does not require a gradient in proportions, is *ridit analysis*. Virtually the only assumption made in ridit analysis is that the discrete categories represent intervals of an underlying but unobservable continuous distribution. No assumption is made about normality or any other form for the distribution.

Ridit analysis is due to Bross (1958), and has been applied to the study of automobile accidents (Bross, 1960), of cancer (Wynder, Bross, and Hirayama, 1960), and of schizophrenia (Spitzer et al., 1965). A mathematical study of ridit analysis was made by Kantor, Winkelstein, and Ibrahim (1968).

Ridit analysis begins with the selection of a population to serve as a standard or reference group. The term ridit is derived from the initials of "relative to an identified distribution." For the reference group, we estimate the proportion of all individuals with a value on the underlying continuum falling at or below the midpoint of each interval, that is, each interval's *ridit.* This initial arithmetic is illustrated in Table 9.7, using data from Bross (1958, p. 20).

Table 9.7.   An illustration of the calculation of ridits for degrees of injury

Severity	(1)	(2)	(3)	(4)	(5) = ridit
None	17	8.5	0	8.5	.047
Minor	54	27.0	17	44.0	.246
Moderate	60	30.0	71	101.0	.564
Severe	19	9.5	131	140.5	.785
Serious	9	4.5	150	154.5	.863
Critical	6	3.0	159	162.0	.905
Fatal	14	7.0	165	172.0	.961

1. In general, column 1 contains the distribution over the various categories for the reference group. In Table 9.7, the distribution is over seven categories of injury for the 179 members of a selected sample.

2. The entries in column 2 are simply half the corresponding entries in column 1.

3. The entries in column 3 are the accumulated entries in column 1, but displaced one category downwards.

4. The entries in column 4 are the sums of the corresponding entries in columns 2 and 3.

5. The entries in column 5, finally, are those in column 4 divided by the total sample size, in this case 179.

The final values are the *ridits* associated with the various categories. The ridit for a category, then, is nothing but the proportion of all subjects from the reference group falling in the lower ranking categories plus half the proportion falling in the given category. If, in the model of an underlying continuum, we assume that the distribution is uniform in each interval, then a category's ridit is the proportion of all subjects from the reference group with an underlying value at or below the midpoint of the corresponding interval.

Given the distribution of any other group over the same categories, the mean ridit for that group may be calculated. The resulting mean value is interpretable as a probability. The mean ridit for a group is the probability that a randomly selected individual from it has a value indicating greater severity or seriousness than a randomly selected individual from the standard group.

In our example, if this probability is .50, we infer that the comparison group tends to sustain neither more nor less serious injuries than the reference group. For the reference group itself, by the way, the mean ridit is necessarily .50. This is consistent with the fact that, if two subjects are randomly selected from the same population, then the second subject will have a more extreme value half the time, and will have a less extreme value also half the time.

If the mean ridit for a comparison group is greater than .50, then more than half of the time a randomly selected subject from it will have a more extreme value than a randomly selected subject from the reference group. In our example, we would infer that the comparison group tends to sustain more serious injuries than the reference group. If, finally, a comparison group's mean ridit is less than .50, we would infer that its subjects tend to have less extreme values than the subjects of the reference group.

As an example, consider the hypothetical data of Table 9.8, giving the distribution of seriousness of injury to the driver when he was involved in an accident and had been slightly intoxicated.

Table 9.8. Seriousness of injury sustained by slightly intoxicated drivers of automobiles involved in accidents

Severity	Number	Ridit	Product
None	5	.047	0.235
Minor	10	.246	2.460
Moderate	16	.564	9.024
Severe	5	.785	3.925
Serious	3	.863	2.589
Critical	6	.905	5.430
Fatal	5	.961	4.805
Total	50		28.468

The mean ridit for a group is simply the sum of the products of observed frequencies times corresponding ridits, divided by the total frequency. For slightly intoxicated drivers the mean is

$$\bar{r} = \frac{28.468}{50} = .57. \tag{9.37}$$

Thus the odds are 4 to 3 $(= .57/.43)$ that a slightly intoxicated driver will sustain a more serious injury than a driver from the reference group if both are involved in accidents.

The standard error of a mean ridit is approximately

$$\text{s.e.}(\bar{r}) = \frac{1}{2\sqrt{3N}}. \tag{9.38}$$

Thus, with $N = 50$, the standard error of the mean ridit just calculated is

$$\text{s.e.}(\bar{r}) = \frac{1}{2\sqrt{150}} = .04. \tag{9.39}$$

The significance of the difference between an obtained mean ridit and the standard value of .5 may be tested by referring the value of

$$z = \frac{\bar{r} - .5}{\text{s.e.}(\bar{r})} \tag{9.40}$$

to Table A.2 of the normal distribution. For our example,

$$z = \frac{.57 - .50}{.04} = 1.75. \tag{9.41}$$

Because z failed to reach significance, we would have to conclude that the seriousness of injuries to slightly intoxicated drivers might equal that to members of the reference group.

As another example of the use of ridit analysis, suppose we have data on a sample of 50 extremely intoxicated drivers who were involved in accidents, and suppose that their mean ridit is .73. An important comparison is between slightly and extremely intoxicated drivers. Instead of identifying a new reference group, all we need to do is to subtract one of the two mean ridits from the other and add .50. Thus we obtain $(.73 - .57) + .50 = .66$ as the chances that a driver who is extremely intoxicated will sustain a more severe injury than one who is slightly so, when they are involved in accidents.

If one mean ridit is based on N_1 subjects and the other on N_2, the standard error is approximately

$$\text{s.e.}(\bar{r}_2 - \bar{r}_1) = \frac{\sqrt{N_1 + N_2}}{2\sqrt{3N_1N_2}}. \tag{9.42}$$

With $N_1 = N_2 = 50$, the approximate standard error is

$$\text{s.e.}(\bar{r}_2 - \bar{r}_1) = \frac{\sqrt{100}}{2\sqrt{3 \times 50 \times 50}}$$

$$= .06. \tag{9.43}$$

The significance of the difference between $\bar{r}_1$ and $\bar{r}_2$ may be tested by referring the value of

$$z = \frac{\bar{r}_2 - \bar{r}_1}{\text{s.e.}(\bar{r}_2 - \bar{r}_1)} \tag{9.44}$$

to Table A.2. For our example,

$$z = \frac{.73 - .57}{.06} = 2.67, \tag{9.45}$$

which indicates a difference significant at the .01 level. We can therefore infer that extremely intoxicated drivers involved in accidents tend to sustain more serious injuries than slightly intoxicated drivers involved in accidents.

Problem 9.1. Prove the equality of expressions 9.3 and 9.4 for χ^2.

Problem 9.2. The estimate of the slope b is given by (9.17). Prove that its numerator is given by (9.19) and that its denominator is given by (9.20).

Problem 9.3. Three samples of New York mental hospital patients were studied as part of a collaborative project (Cooper et al., 1972). The numbers of hospital diagnoses of affective disorders were as follows:

Sample	Age Range	Number of Patients	Number Diagnosed Affective	Proportion
1	20–34	105	2	
2	20–59	192	13	
3	35–59	145	24	
Overall		442	39	

(*a*) Calculate the proportions diagnosed affective and test whether they differ significantly.

(*b*) Test whether the proportion diagnosed affective in the first two samples combined differs significantly from the proportion in the third sample. Test whether the proportions in the first two samples differ.

(*c*) The patients in sample 1 tend to be younger than those in sample 2, who in turn tend to be younger than those in sample 3. Because the chances of an affective disorder increase with age, it might be hypothesized that p_1 should be less than p_2, and that p_2 in turn should be less than p_3. Are the proportions in this order? What is the value of $\bar{\chi}^2$ (9.32)? What is the value of c (9.34)? Refer to Table A.6 to test the hypothesized ordering.

Problem 9.4. Verify that the mean ridit in Table 9.7 is .50.

REFERENCES

Armitage, P. (1955). Tests for linear trends in proportions and frequencies. *Biometrics*, **11**, 375–385.

Barlow, R. E., Bartholomew, D. J., Bremner, J. M. and Brunk, H. D. (1972). *Statistical inference under order restrictions*. New York: Wiley.

Bartholomew, D. J. (1959a). A test of homogeneity for ordered alternatives. *Biometrika*, **46**, 36–48.

Bartholomew, D. J. (1959b). A test of homogeneity for ordered alternatives II. *Biometrika*, **46**, 328–335.

Bartholomew, D. J. (1963). On Chassan's test for order. *Biometrics*, **19**, 188–191.

Bross, I. D. J. (1958). How to use ridit analysis. *Biometrics*, **14**, 18–38.

Bross, I. D. J. (1960). How to cut the highway toll in half in the next ten years. *Public Health Rep.*, **75**, 573–581.

Castellan, N. J. (1965). On the partitioning of contingency tables. *Psychol. Bull.*, **64**, 330–338.

Chapman, D. G. and Nam, J. (1968). Asymptotic power of chi square tests for linear trends in proportions. *Biometrics*, **24**, 315–327.

Chassan, J. B. (1960). On a test for order. *Biometrics*, **16**, 119–121.

Chassan, J. B. (1962). An extension of a test for order. *Biometrics*, **18**, 245–247.

Cochran, W. G. (1954). Some methods of strengthening the common χ^2 tests. *Biometrics*, **10**, 417–451.

Cooper, J. E., Kendell, R. E., Gurland, B. J., Sharpe, L., Copeland, J. R. M. and Simon, R. (1972). *Psychiatric diagnosis in New York and London*. London: Oxford University Press.

Dorn, H. F. (1954). The relationship of cancer of the lung and the use of tobacco. *Amer. Statist.*, **8**, 7–13.

Irwin, J. O. (1949). A note on the subdivision of chi-square into components. *Biometrika*, **36**, 130–134.

Kantor, S., Winkelstein, W. and Ibrahim, M. A. (1968). A note on the interpretation of the ridit as a quantile rank. *Amer. J. Epidemiol.*, **87**, 609–615.

Kastenbaum, M. A. (1960). A note on the additive partitioning of chi-square in contingency tables. *Biometrics*, **16**, 416–422.

Kimball, A. W. (1954). Short-cut formulas for the exact partition of χ^2 in contingency tables. *Biometrics*, **10**, 452–458.

Lancaster, H. O. (1950). The exact partition of chi-square and its application to the problem of pooling small expectations. *Biometrika*, **37**, 267–270.

Mantel, N. (1963). Chi-square tests with one degree of freedom: Extensions of the Mantel-Haenszel procedure. *J. Amer. statist. Assoc.*, **58**, 690–700.

Miller, R. G. (1966). *Simultaneous statistical inference*. New York: McGraw-Hill.

National Center for Health Statistics (1970). Selected symptoms of psychological distress in the United States. *Data from National Health Survey*, Series 11, No. 37.

Snell, E. J. (1964). A scaling procedure for ordered categorical data. *Biometrics*, **20**, 592–607.

Spitzer, R. L., Fleiss, J. L., Kernohan, W., Lee, J. and Baldwin, I. T. (1965). The Mental Status Schedule: Comparing Kentucky and New York schizophrenics. *Arch. gener. Psychiatr.*, **12**, 448–455.

Wynder, E. L., Bross, I. D. J. and Hirayama, T. (1960). A study of the epidemiology of cancer of the breast. *Cancer*, **13**, 559–601.

Yates, F. (1948). The analysis of contingency tables with groupings based on quantitative characteristics. *Biometrika*, **35**, 176–181.

CHAPTER 10

Combining Evidence from Fourfold Tables

If the possibility of an association between a factor A and a disease B is strong, interesting, or important enough, it is virtually guaranteed that a number of investigators will study the association. Similarly, if an association has been found to exist in one kind of population, it is to be expected that the possibility of association in other kinds of populations will be studied.

Suppose that the association between A and B has been studied in each of g groups, with each group generating its own fourfold table. The following questions can be asked:

1. Is there evidence that the degree of association, whatever its magnitude, is consistent from one group to another?
2. Assuming that the degree of association is found to be consistent, is the common degree of association statistically significant?
3. Assuming that the common degree of association is significant, what is the best estimate of the common value for the measure of association? What is its standard error?

Section 10.1 provides a mathematical framework within which these questions can be answered. Section 10.2 describes a method due to Cochran, Section 10.3 describes a method using the logarithm of the odds ratio, and Section 10.4 describes the Mantel-Haenszel method. Section 10.5 indicates how these methods can be used as alternatives to matching in the control of confounding factors. Section 10.6 describes some of the more popular but generally invalid methods for combining data from fourfold tables. Some references to the more general analysis of multiple contingency tables are given in Section 10.7.

10.1. SOME THEORY OF CHI SQUARE

To answer the questions posed above, some knowledge of the theory of chi square tests is necessary. For a single one of the g groups, say the ith, let

m_i denote the value of the chosen measure of association. The measure might be the difference between two proportions, the logarithm of the odds ratio, and so on.

Whatever m_i is, let s.e.(m_i) denote its standard error and define

$$w_i = \frac{1}{(\text{s.e.}(m_i))^2}, \tag{10.1}$$

so that w_i is the reciprocal of the squared standard error of m_i. The quantity w_i is the weight to be attached to m_i. If the standard error of m_i is large, implying that m_i is not too precise, then w_i is small. This is reasonable, since imprecise estimates should not be given great weight. If, on the other hand, the standard error of m_i is small, implying that m_i is rather precise, then w_i is large. This, too, is reasonable, since precise estimates should be given great weight.

Let us suppose that m_i is such that a value of zero indicates no association. Then, when the hypothesis of no association in the ith group is true, the quantity

$$\chi_i = \frac{m_i}{\text{s.e.}(m_i)} = m_i\sqrt{w_i} \tag{10.2}$$

has, approximately, the standard normal distribution, and the quantity

$$\chi_i^2 = w_i m_i^2 \tag{10.3}$$

has, approximately, the chi square distribution with one degree of freedom. If the hypothesis of no association in the ith group is false, χ_i^2 may be expected to be large, so that the hypothesis is likely to be rejected if a chi square test is applied.

We are not so much interested in the ith group, or in any other single group, however, as in all the groups together. The analysis of all groups conveniently begins with the calculation of

$$\chi_{\text{total}}^2 = \sum_{i=1}^{g} \chi_i^2 = \sum_{i=1}^{g} w_i m_i^2. \tag{10.4}$$

If there is no association in any of the g groups, then χ_{total}^2 has a chi square distribution with g degrees of freedom. This follows because the sum of g independent chi squares, each with one degree of freedom, has a chi square distribution with g degrees of freedom and because the g groups are assumed to be independent.

If we calculate χ_{total}^2 and find it to be significantly large, we may validly conclude that there is association somewhere within the g groups. We would not, however, know whether the association was consistent across all groups or whether it varied from one group to another. χ_{total}^2 is not, therefore, informative by itself. Rather, its calculation serves the purpose of simplifying other calculations, as will now be indicated.

χ^2_{total} is subdivided, or partitioned, into two components,

$$\chi^2_{\text{total}} = \chi^2_{\text{homog}} + \chi^2_{\text{assoc}}.$$ (10.5)

The quantity χ^2_{homog} assesses the degree of homogeneity, or equality, among the g measures of association, and the quantity χ^2_{assoc} assesses the significance of the average degree of association. The subdivision indicated by (10.5) is most easily effected by calculating χ^2_{assoc} first, and determining χ^2_{homog} by simple subtraction.

The term χ^2_{assoc} is calculated as follows. The overall measure of association across all groups is taken as the weighted average of the g individual measures, with the weights being those defined in (10.1):

$$\bar{m} = \frac{\sum\limits_{i=1}^{g} w_i m_i}{\sum\limits_{i=1}^{g} w_i}.$$ (10.6)

Under the hypothesis that the overall association is zero, $\bar{m}$ has an average value of zero and a standard error of

$$\text{s.e.}(\bar{m}) = \frac{1}{\sqrt{\sum\limits_{i=1}^{g} w_i}}.$$ (10.7)

Hence

$$\chi_{\text{assoc}} = \frac{\bar{m}}{\text{s.e.}(\bar{m})} = \frac{\sum\limits_{i=1}^{g} w_i m_i}{\sqrt{\sum\limits_{i=1}^{g} w_i}}$$ (10.8)

is distributed approximately as a standard normal variate under the hypothesis, and

$$\chi^2_{\text{assoc}} = \bar{m}^2 \sum\limits_{i=1}^{g} w_i = \frac{\left(\sum\limits_{i=1}^{g} w_i m_i\right)^2}{\sum\limits_{i=1}^{g} w_i}$$ (10.9)

is distributed approximately as chi square with one degree of freedom.

The term χ^2_{homog} is then easily obtained by subtraction:

$$\chi^2_{\text{homog}} = \chi^2_{\text{total}} - \chi^2_{\text{assoc}} = \sum\limits_{i=1}^{g} w_i m_i^2 - \bar{m}^2 \sum\limits_{i=1}^{g} w_i.$$ (10.10)

An equivalent expression for χ^2_{homog} is

$$\chi^2_{\text{homog}} = \sum_{i=1}^{g} w_i(m_i - \bar{m})^2. \tag{10.11}$$

This expression for χ^2_{homog} is useful for two purposes. One is to provide a numerical check on the arithmetic. The other is to point out that χ^2_{homog} actually measures the degree of variability among the separate values of m_i. χ^2_{homog} is approximately distributed as chi square with $g - 1$ degrees of freedom under the hypothesis of consistent (homogeneous) association.

Means are therefore provided for answering the three questions posed at the beginning of this chapter.

1. Consistency of association can be tested by referring χ^2_{homog} to tables of chi square with $g - 1$ degrees of freedom. If χ^2_{homog} is significant, the next step in the analysis would be to partition χ^2_{homog} into appropriate components in order to identify those groups in which the association is different from that in the remaining groups (see Sections 8.4 and 9.1, and Problem 10.1 (c)–(e)).

2. If χ^2_{homog} is not significant, the significance of the overall degree of association can be tested by referring χ^2_{assoc} to tables of chi square with one degree of freedom.

3. The best estimate of the overall degree of association is $\bar{m}$ (10.6). Its standard error is given by (10.7).

What remains, then, is to apply these results to particular choices of the measure of association. The following notation will be used consistently in this chapter. In the ith group, n_{i1} is the number of observations in the first sample and p_{i1} is the proportion of the first sample having the studied characteristic. The quantity n_{i2} is the number of observations in the second sample, and p_{i2} is the proportion of the second sample having the studied characteristic. The total number of observations in the ith group is denoted by $n_{i.} = n_{i1} + n_{i2}$, and the overall proportion in the ith group having the characteristic is denoted by

$$\bar{p}_i = \frac{n_{i1}p_{i1} + n_{i2}p_{i2}}{n_{i.}}. \tag{10.12}$$

The complementary proportion is $\bar{q}_i = 1 - \bar{p}_i$.

10.2. COCHRAN'S METHOD

A number of measures of association were introduced in Chapter 5. Another measure which has much to recommend it (Yates, 1955; Fleiss,

1970) is the so-called *standardized difference*,

$$d = \frac{p_1 - p_2}{\bar{p}\bar{q}} . \tag{10.13}$$

Cochran's method (1954) for combining evidence from a number of fourfold tables fits the mold of Section 10.1, with the measure of association for the ith group being its standardized difference,

$$m_i = d_i = \frac{p_{i1} - p_{i2}}{\bar{p}_i\bar{q}_i} . \tag{10.14}$$

The squared standard error is

$$(\text{s.e.}(d_i))^2 = \frac{1}{\bar{p}_i\bar{q}_i}\left(\frac{n_{i.}}{n_{i1}n_{i2}}\right), \tag{10.15}$$

so that

$$w_i = \frac{\bar{p}_i\bar{q}_i n_{i1} n_{i2}}{n_{i.}} \tag{10.16}$$

and

$$\chi_i^2 = w_i d_i^2 = \frac{(p_{i1} - p_{i2})^2}{\bar{p}_i\bar{q}_i(1/n_{i1} + 1/n_{i2})}, \tag{10.17}$$

precisely the usual chi square value, aside from the continuity correction.

Table 10.1 presents the proportions of patients diagnosed as schizophrenic by a cooperative team of psychiatrists in New York and London (see Cooper et al., 1972).

Table 10.1. Diagnoses of schizophrenia by a cooperative project team in three studies in New York and London

Study	New York		London	
	n_{i1}	p_{i1}	n_{i2}	p_{i2}
$i = 1$ (ages 20–34)	105	.362	105	.314
$i = 2$ (ages 20–59)	192	.292	174	.351
$i = 3$ (ages 35–59)	145	.297	145	.228

Table 10.2 outlines the arithmetic required to perform Cochran's analysis.

Table 10.2. Cochran's method applied to data of Table 10.1

Study	$p_{i1} - p_{i2}$	$\bar{p}_i$	d_i	$n_{i.}$	w_i	$w_i d_i$	$w_i d_i^2$
1	.048	.338	.215	210	11.75	2.53	0.54
2	−.059	.320	−.271	366	19.86	−5.38	1.46
3	.069	.263	.356	290	14.05	5.00	1.78
Total					45.66	2.15	3.78

None of the individual chi squares ($= w_i d_i^2$) is significant, nor is the total chi square with three degrees of freedom,

$$\chi^2_{\text{total}} = \sum_{i=1}^{3} w_i d_i^2 = 3.78. \tag{10.18}$$

The value of the chi square statistic for testing the homogeneity of the standardized differences is

$$\chi^2_{\text{homog}} = \sum_{i=1}^{3} w_i d_i^2 - \frac{\left(\sum_{i=1}^{3} w_i d_i\right)^2}{\sum_{i=1}^{3} w_i}$$

$$= 3.78 - \frac{(2.15)^2}{45.66}$$

$$= 3.68 \tag{10.19}$$

with two degrees of freedom, clearly not significant.

The differences between New York and London in the proportions diagnosed schizophrenic by the project therefore seem to be consistent across the three studies. The estimate of the apparently common standardized difference is

$$\bar{d} = \frac{\sum_{i=1}^{3} w_i d_i}{\sum_{i=1}^{3} w_i} = \frac{2.15}{45.66}$$

$$= .047, \tag{10.20}$$

with an estimated standard error of

$$\text{s.e.}(\bar{d}) = \frac{1}{\sqrt{\sum_{i=1}^{3} w_i}} = \frac{1}{\sqrt{45.66}}$$

$$= .15. \tag{10.21}$$

The value of the chi square statistic for testing the significance of this mean standardized difference is

$$\chi^2_{\text{assoc}} = \left(\frac{\bar{d}}{\text{s.e.}(\bar{d})}\right)^2 = \left(\frac{.047}{.15}\right)^2 = 0.10 \tag{10.22}$$

with one degree of freedom. The conclusion would then be that, based on the project team's diagnosis, there is no difference between New York and London mental hospital patients in the proportions diagnosed schizophrenic.

Cochran's method of analysis has been shown by Radhakrishna (1965) to be powerful under rather general conditions. Its major drawback is the relative unfamiliarity of the standardized difference as a measure of association.

10.3. COMBINING THE LOGARITHMS OF ODDS RATIOS

The odds ratio itself,

$$o_i = \frac{p_{i1}(1 - p_{i2})}{p_{i2}(1 - p_{i1})}, \tag{10.23}$$

does not have the property that a value of zero indicates no association. The logarithm of the odds ratio does have this property. Thus, consider taking as the measure of association

$$m_i = L_i = \log_e (o_i). \tag{10.24}$$

The squared standard error of L_i is approximately

$$(\text{s.e.}(L_i))^2 = \frac{1}{w_i} = \frac{1}{n_{i1}p_{i1}(1 - p_{i1})} + \frac{1}{n_{i2}p_{i2}(1 - p_{i2})}, \tag{10.25}$$

which is equal to the sum of the reciprocals of the frequencies within the cells. The weight w_i is then the reciprocal of this sum of reciprocals.

The chi square analyses of Section 10.1 have been applied to the logarithm of the odds ratio by Gart (1962) and Sheehe (1966), and will now be illustrated on the data of Table 10.1.

Table 10.3. Analysis of logarithms of odds ratios
applied to data of Table 10.1

Study	o_i	L_i	w_i	w_iL_i	$w_iL_i^2$
1	1.24	0.215	11.71	2.52	0.54
2	0.76	−0.274	19.82	−5.43	1.49
3	1.43	0.358	13.83	4.95	1.77
Total			45.36	2.04	3.80

The individual chi squares ($= w_iL_i^2$) are all approximately equal to those in Table 10.2. When the association in a study is strong, however, the value of chi square based on the logarithm of the odds ratio may be expected to be less than the value of the standard chi square (see Problem 10.1).

The value of the total chi square is

$$\chi^2_{\text{total}} = \sum_{i=1}^{3} w_iL_i^2 = 3.80, \tag{10.26}$$

only trivially greater than the value given by (10.18) for Table 10.2. The value of the chi square statistic for testing the homogeneity of the odds ratios is

$$\chi^2_{\text{homog}} = \sum_{i=1}^{3} w_i L_i^2 - \frac{\left(\sum_{i=1}^{3} w_i L_i\right)^2}{\sum_{i=1}^{3} w_i}$$

$$= 3.80 - \frac{(2.04)^2}{45.36}$$

$$= 3.71 \tag{10.27}$$

with two degrees of freedom, indicating no significant differences among the three odds ratios. This value of χ^2_{homog} is close to the value given by (10.19) for the data of Table 10.2.

The estimate of the logarithm of the common odds ratio is

$$L = \frac{\sum_{i=1}^{3} w_i L_i}{\sum_{i=1}^{3} w_i} = \frac{2.04}{45.36}$$

$$= 0.045, \tag{10.28}$$

with an estimated standard error of

$$\text{s.e.}(L) = \frac{1}{\sqrt{\sum_{i=1}^{3} w_i}} = \frac{1}{\sqrt{45.36}}$$

$$= 0.15, \tag{10.29}$$

yielding a value for chi square of

$$\chi^2_{\text{assoc}} = \left(\frac{L}{\text{s.e.}(L)}\right)^2 = \left(\frac{.045}{.15}\right)^2 = 0.09 \tag{10.30}$$

with one degree of freedom. The inference drawn is the same as that based on (10.22) for the standardized difference, namely, that there is no difference between New York and London mental hospital patients in the proportions diagnosed schizophrenic by the project. The mean odds ratio, a more familiar measure than L, is estimated by

$$\bar{o} = e^L = \text{antilog}(L), \tag{10.31}$$

with an approximate standard error of

$$\text{s.e.}(\bar{o}) = \bar{o} \cdot \text{s.e.}(L). \tag{10.32}$$

For our data, the mean odds ratio is

$$\bar{o} = e^{.045} = 1.05, \qquad (10.33)$$

so that the odds of a New York patient's being diagnosed schizophrenic by the project are effectively equal to the corresponding odds for a London patient. The standard error of $\bar{o}$ is approximately

$$\text{s.e.}(\bar{o}) = 1.05 \times 0.15 = 0.16. \qquad (10.34)$$

Naylor (1967) and Gart (1970, 1971) have shown that the estimates of the mean odds ratio and its standard error may be biased if the sample sizes are small. Accuracy is improved by the addition of .5 to each cell frequency as in (5.20) and (5.33).

10.4. THE MANTEL-HAENSZEL METHOD

A method due to Mantel and Haenszel (1959) and extended by Mantel (1963) may be viewed as combining some of the advantages of the methods described in the two preceding sections. The chi square statistics associated with it are virtually identical to those associated with the Cochran method. The only two modifications are in taking

$$d_i = \frac{n_{i.} - 1}{n_{i.}} \frac{p_{i1} - p_{i2}}{\bar{p}_i \bar{q}_i} \qquad (10.35)$$

and

$$w_i = \frac{\bar{p}_i \bar{q}_i n_{i1} n_{i2}}{n_{i.} - 1}. \qquad (10.36)$$

Expressions (10.35) and (10.36) differ only trivially from (10.14) and (10.16) if $n_{i.}$, the total sample size in the ith group, is at all large, but may differ appreciably if $n_{i.}$ is small. In the latter case, (10.35) and (10.36) are preferable.

Table 10.4 compares the values of d_i and of w_i used in the Cochran and in the Mantel-Haenszel methods. The differences are seen to be only slight.

Table 10.4. *Values of d_i and w_i used in the Cochran and Mantel-Haenszel methods for data of Table 10.1*

	d_i		w_i	
Study	Cochran	Mantel-Haenszel	Cochran	Mantel-Haenszel
1	.215	.214	11.75	11.80
2	−.271	−.270	19.86	19.91
3	.356	.355	14.05	14.09

The Mantel-Haenszel method yields a summary estimate of the odds ratio, whereas the Cochran does not. The estimate, not requiring logarithms, is

$$\hat{o} = \frac{\sum_{i=1}^{g}(n_{i1}n_{i2}/n_{i.})p_{i1}(1 - p_{i2})}{\sum_{i=1}^{g}(n_{i1}n_{i2}/n_{i.})p_{i2}(1 - p_{i1})}. \tag{10.37}$$

Table 10.5 outlines the required arithmetic.

Table 10.5.   Calculation of Mantel-Haenszel estimate of mean odds ratio for data of Table 10.1

Study	(1) $n_{i1}n_{i2}/n_{i.}$	(2) $p_{i1}(1 - p_{i2})$	(3) $p_{i2}(1 - p_{i1})$	(4) (1) × (2)	(5) (1) × (3)
1	52.5	.248	.200	13.02	10.50
2	91.3	.190	.249	17.35	22.73
3	72.5	.229	.160	16.60	11.60
Total				46.97	44.83

The mean odds ratio is the ratio of the sum in column 4 to that in column 5,

$$\hat{o} = \frac{46.97}{44.83} = 1.05, \tag{10.38}$$

which happens to equal the estimate given by (10.33). In general, $\hat{o}$ (10.37) is different from and more accurate than $\bar{o}$ (see Gart, 1970). The only disadvantage to the Mantel-Haenszel method appears to be that a simple expression for the standard error of $\hat{o}$ does not exist.

10.5.   AN ALTERNATIVE TO MATCHING

The methods of Sections 10.2 to 10.4 can be used as alternatives to the matching of subjects in comparative retrospective and prospective studies. Suppose that a retrospective study is contemplated of the association between cigarette smoking and lung cancer, with control for the possible confounding effects of age and sex. One method of control is to pair each lung cancer case with one or more controls of the same sex and of a similar age, and to apply the methods of Section 8.1 or 8.3.

Another method of control is to draw a cross-sectional sample of cases and a cross-sectional sample of controls, to stratify the two samples by sex and by age intervals, and then, separately for each resulting stratum, to set up a fourfold table contrasting the rates of smoking for the cases and controls. If there are, say, five age intervals, the total number of fourfold tables is $g = 10$: five for the males plus five for the females. The resulting set of tables may be viewed as coming from g distinct groups.

Matching has the advantage of assuring that the two samples are comparable on the factors used for matching, but has as a major disadvantage the practical difficulty of finding a matched control for each case if the number of cases is large. Other disadvantages are cited in Section 8.5.

Stratifying the samples after they have been drawn has the advantage of not requiring a specification beforehand of the composition of the two samples, as well as the advantage of permitting an examination of the consistency of association across the various strata. A disadvantage is that, if the sample sizes are not large, the number of individuals in a stratum from one sample may be small compared to the number of individuals in it from the other sample. The power and precision of the comparisons may therefore suffer.

Research on the effectiveness of matching versus the effectiveness of stratification in controlling for confounding factors has been performed by Cochran (1968) and Rubin (1970) for quantitative measurements. To the extent that their results can be applied to the comparison of proportions, we may view matching as the method of choice for small sample sizes and cross-sectional sampling followed by stratification as the method of choice for large sample sizes.

10.6. METHODS TO BE AVOIDED

The Summation of Chi Procedure

One of the more frequently employed methods for combining data from different fourfold tables is of the form outlined in Section 10.1, although not obviously so. The method, usually referred to as the summation of chi procedure, has long been known to have serious defects but keeps reappearing nevertheless (see, e.g., Finney, 1965). The method in effect takes as the measure of association

$$m_i = z_i = \frac{p_{i1} - p_{i2}}{\sqrt{\bar{p}_i \bar{q}_i (1/n_{i1} + 1/n_{i2})}}.$$ (10.39)

Because z_i has been standardized to have a standard error of unity, therefore

$$w_i = \frac{1}{(\text{s.e.}(z_i))^2} = 1.$$ (10.40)

The word "chi" in the name of the procedure derives from z_i's being the square root of a chi square variate (without the correction for continuity), and hence being a *chi* variate.

When m_i is defined by (10.39), the value of the total chi square (10.4) is given by

$$\chi^2_{\text{total}} = \sum_{i=1}^{g} z_i^2. \tag{10.41}$$

Furthermore,

$$\bar{m} = \frac{\sum_{i=1}^{g} z_i}{g} = \bar{z} \tag{10.42}$$

and

$$\sum_{i=1}^{g} w_i = g, \tag{10.43}$$

by (10.40). Therefore, by (10.9),

$$\chi^2_{\text{assoc}} = \frac{\left(\sum_{i=1}^{g} z_i \right)^2}{g} = g\bar{z}^2, \tag{10.44}$$

and, by (10.11),

$$\chi^2_{\text{homog}} = \sum_{i=1}^{g} (z_i - \bar{z})^2. \tag{10.45}$$

There are serious flaws inherent in the interpretation of the latter two chi squares (see, e.g., Pasternack and Mantel, 1966). Consider the numerical example of Tables 10.6 and 10.7.

Table 10.6.   Association between A and B in group 1

	B	$\bar{B}$	Total
A	60	40	100
$\bar{A}$	40	60	100
Total	100	100	200

For group 1 (Table 10.6),

$$z_1 = \frac{.6 - .4}{\sqrt{.5 \times .5(\frac{1}{100} + \frac{1}{100})}} = 2.83, \tag{10.46}$$

and

$$\chi_1^2 = z_1^2 = 8.00. \tag{10.47}$$

Suppose that, in group 2, the relative frequencies are precisely the same as in group 1, but that samples ten times as large are studied (Table 10.7).

Table 10.7. Association between A and B in group 2

	B	$\bar{B}$	Total
A	600	400	1000
$\bar{A}$	400	600	1000
Total	1000	1000	2000

For group 2,

$$z_2 = \frac{.6 - .4}{\sqrt{.5 \times .5(\frac{1}{1000} + \frac{1}{1000})}} = 8.90, \tag{10.48}$$

and

$$\chi_2^2 = z_2^2 = 80. \tag{10.49}$$

The average value of z is

$$\bar{z} = \tfrac{1}{2}(2.83 + 8.90) = 5.86, \tag{10.50}$$

so that, by (10.44),

$$\chi_{\text{assoc}}^2 = 2 \times (5.86)^2 = 68.80 \tag{10.51}$$

with one degree of freedom. What is disquieting about this value for the overall test of association is that it is less than the value for one of the individual chi squares for association, $\chi_2^2 = 80$. The addition of the evidence from group 1, in which the association was really the same as in group 2, would be expected to increase the statistical significance of the association. The summation of chi procedure failed to do so.

The value of the chi square statistic for homogeneity is

$$\chi_{\text{homog}}^2 = \chi_{\text{total}}^2 - \chi_{\text{assoc}}^2$$
$$= (8 + 80) - 68.8 = 19.2, \tag{10.52}$$

also with one degree of freedom. The inference would be unequal degrees of association in the two groups, in spite of the identity of the proportions in them. What χ_{homog}^2 is measuring, in this instance, is the difference between the sample sizes studied from the two groups rather than any difference in association.

Any procedure for which an accumulation of evidence for association may lead to a reduction in chi square, and for which inequalities in sample sizes may contribute to the inference of heterogeneous association, is to be avoided. So be it with the summation of chi procedure.

Summation Observed Versus Summation Expected

A relative lack of sensitivity to added evidence for association in a given direction characterizes the following method, too. Although it can be cast into the terms of the theory of Section 10.1, it would not be an aid to under-standing.

The method calls first for generating a total fourfold table by summing the frequencies across the g individual tables. Let the observed frequencies for $g = 2$ groups be as in Table 10.8. Chi square is calculated without the continuity correction. The association between A and B is clearly the same

Table 10.8.   Association between A and B in two groups

	Group 1				Group 2		
	B	$\bar{B}$	Total		B	$\bar{B}$	Total
A	300	200	500	A	300	200	500
$\bar{A}$	200	300	500	$\bar{A}$	400	600	1000
Total	500	500	1000	Total	700	800	1500
	$\chi_1^2 = 40.0$				$\chi_2^2 = 53.1$		

in the two groups, the only difference between them being that twice as many $\bar{A}$ subjects are studied in group 2 as in group 1. The total chi square is then

$$\chi_{\text{total}}^2 = 40.0 + 53.1 = 93.1. \tag{10.53}$$

The table of total frequencies is as shown in Table 10.9.

**Table 10. 9.   Sum of frequencies
for groups 1 and 2**

	B	$\bar{B}$	Total
A	600	400	1000
$\bar{A}$	600	900	1500
Total	1200	1300	2500

The next step is to determine, for each group, the set of frequencies expected under the hypothesis of no association. The expected frequency per cell is calculated as the product of the total frequencies in a cell's row and column, divided by the overall frequency in the table. Thus the expected frequency in the (A, B) cell for group 2 is $500 \times 700/1500 = 350,000/1500 = 233$. The expected frequencies are as in Table 10.10.

Next, generate an overall table of expected frequencies by summing across the individual tables just determined (Table 10.11).

Table 10.10. Expected frequencies for groups 1 and 2

	Group 1				Group 2		
	B	$\bar{B}$	Total		B	$\bar{B}$	Total
A	250	250	500	A	233	267	500
$\bar{A}$	250	250	500	$\bar{A}$	467	533	1000
Total	500	500	1000	Total	700	800	1500

Table 10.11. Sum of expected frequencies for groups 1 and 2

	B	$\bar{B}$	Total
A	483	517	1000
$\bar{A}$	717	783	1500
Total	1200	1300	2500

Finally, calculate the summary chi square for association by taking, for each cell, the difference between the total observed and total expected frequencies, squaring, dividing by the total expected frequency, and summing across all four cells. Thus, from Tables 10.9 and 10.11,

$$\chi^2_{assoc} = \frac{(600 - 483)^2}{483} + \frac{(400 - 517)^2}{517} + \frac{(600 - 717)^2}{717} + \frac{(900 - 783)^2}{783}$$

$$= 80.8. \tag{10.54}$$

Unlike the chi square for association generated by the summation of chi procedure, this chi square is larger than either of the two individual chi squares. Nevertheless, it is not quite large enough. The value of the chi square statistic for testing the equality of association in the two groups is the difference between (10.53) and (10.54):

$$\chi^2_{homog} = 93.1 - 80.8 = 12.3, \tag{10.55}$$

which is highly significant. Since the association between A and B is identical in the two groups, it is clear that this method, as the summation of chi procedure, is sensitive to irrelevant differences in sample sizes. Therefore, this method is also to be avoided.

Chi Square on the Table of Totals

A defect opposite in nature to that of the two preceding methods characterizes the following procedure for testing overall association. The method cannot be described in terms of the theory of Section 10.1. It calls merely for generating the table of total observed frequencies as described in the preceding section, and then calculating a straightforward chi square on it.

The method works quite well on the data of the two previous sections, and in general for groups in which corresponding proportions are nearly equal. Such a state of affairs is exceptional, however. Consider the data of Table 10.12. No association between A and B exists in either group, although the basic rates are different in the two groups.

Table 10.12.   Association between A and B in two groups

	Group 1				Group 2		
	B	$\bar{B}$	Total		B	$\bar{B}$	Total
A	10	40	50	A	60	40	100
$\bar{A}$	20	80	100	$\bar{A}$	30	20	50
Total	30	120	150	Total	90	60	150
	$\chi_1^2 = 0$				$\chi_2^2 = 0$		

The table of total frequencies is as shown in Table 10.13. Its associated chi square is 5.01, indicating an association significant at the .05 level. The

Table 10.13.   Sum of frequencies for groups 1 and 2

	B	$\bar{B}$	Total
A	70	80	150
$\bar{A}$	50	100	150
Total	120	180	300

combination of data from tables with unequal proportions and with unequal ratios of sample sizes, n_{i1}/n_{i2}, has created the impression of association where none basically existed.

The procedures just described should be avoided for the reasons indicated (see also Gart, 1962, and Sheehe, 1966). This necessarily means that the calculations become more complicated, as was seen in Sections 10.2 to 10.4, but this is the price one must pay for a valid analysis.

10.7.   MULTIPLE CONTINGENCY TABLES

The problem so far studied in this chapter, that of making inferences about association in a number of fourfold tables, is a special case of the analysis of data from multiple contingency tables, that is, of frequencies cross-classified by more than two characteristics.

The case that seems to have received the greatest amount of attention in the literature is when one of the characteristics represents a dichotomous response or outcome and the other characteristics represent treatments or classificatory variables. Corresponding to each combination of treatments or of classificatory variables, then, there exists the probability that a subject with each of the specified characteristics will exhibit the response or outcome under study. The problem is to test hypotheses about the effects on these probabilities of the treatments or classificatory variables taken singly and in combination, and to estimate these effects.

The mathematical model most frequently assumed is the *logistic model* (see Section 5.4). If P is a typical probability, then the logistic model postulates that the *logit* of P,

$$\text{logit } (P) = \log_e \left(\frac{P}{1 - P}\right), \tag{10.56}$$

is a linear function of the effects of the various factors taken singly and in combination.

Bartlett (1935) seems to have been the first to analyze data from multiple contingency tables using the logistic model. Bartlett's method of analysis was simplified, but only slightly, by Norton (1945). Winsor (1948) fitted a linear model to the logits of the various probabilities when each of the several factors was a dichotomous characteristic. Dyke and Patterson (1952) generalized the analysis to the case where the factors had more than two categories.

Other methods of estimation and of hypothesis testing within the context of a logistic model have been given by Grizzle (1961, 1963) and Berkson (1968). Summary articles have been written by Cox (1958) and Gart (1971), and a text by Cox (1970). The description of a general computer program for fitting a logistic model is given by Lewis (1968). Worked examples illustrating the complicated arithmetic necessary in fitting a logistic model are given by Maxwell (1961, Chapter 6) and by Maxwell and Everitt (1970).

Other models than the logistic have been assumed for proportions. Tallis (1964) has assumed a linear model for the proportions themselves; Mantel (1966) and Berry (1970) have assumed a linear model for the logarithms of the proportions; and Birch (1963) has assumed a linear model for the logarithms of the original *frequencies*. Bishop (1969) has shown Birch's model to be the most general.

Methods for analyzing data when three factors, each with more than two categories, are cross-classified have been suggested by Kastenbaum and Lamphiear (1959) and by Goodman (1963, 1964). Bhapkar and Koch

(1968a, 1968b) and Goodman (1971) have emphasized that the proper inferences from such analyses depend on whether the frequencies were generated according to sampling method I (Chapter 5), method II (Chapter 6), method III (Chapter 7), or to combinations of these methods.

The following is an example of a combination of all three methods of sampling. A specified number of male patients and a specified number of female patients are selected (method II), the selected patients of each sex are distributed into 10-year age intervals (method I), and, within each age interval and separately for each sex, treatments are assigned to patients at random (method III). Response to treatment forms the fourth criterion of classification.

Hypotheses to be tested include whether the treatments have differential effects on the responses, whether the effects of the treatments are the same for males as for females, and whether the responses vary with age. The mathematics required to perform the appropriate analyses is beyond the scope of this book. The reader is referred to the cited articles for computational methods.

Problem 10.1. Table 10.1 gave the proportions of patients in three studies diagnosed schizophrenic by a team of psychiatrists. The following table gives the proportions diagnosed schizophrenic by the hospital psychiatrists.

| Study | New York | | London | |
	n_{i1}	p_{i1}	n_{i2}	p_{i2}
1	105	.771	105	.324
2	192	.615	174	.397
3	145	.566	145	.359

(a) Apply Cochran's method of analysis (Section 10.2) to these data. What inferences would you draw about differences between New York and London hospital psychiatrists in the frequencies with which they diagnose schizophrenia?

(b) Apply the methods of Section 10.3 to these data. What inferences would you draw? How do the values of the chi square statistics compare with those calculated in (a)? What does this comparison say about the powers of the two approaches?

(c) Refer to your results in part (a). Are the standardized differences between New York and London homogeneous? The standardized differences d_2 and d_3 appear to be similar. Test whether they differ significantly. (*Hint.* Refer the value of

$$\chi^2_{2\,vs\,3} = \frac{w_2 w_3}{w_2 + w_3} (d_2 - d_3)^2$$

to critical values of chi square with two degrees of freedom rather than with one because the comparison was suggested by the data.)

(d) The standardized differences d_2 and d_3 differ by less than either does from d_1. Test whether the mean of d_2 and d_3 differs significantly from d_1. (Hint. The mean of d_2 and d_3 is $\bar{d}_{2,3} = (w_2 d_2 + w_3 d_3)/(w_2 + w_3)$. Refer the value of

$$\chi^2_{1\,vs\,(2,3)} = \frac{w_1(w_2 + w_3)}{w_1 + w_2 + w_3}(d_1 - \bar{d}_{2,3})^2$$

to critical values of chi square with two degrees of freedom.)

(e) How does the sum of the chi square statistics determined in (c) and (d) compare with the value of χ^2_{homog} determined in (a)?

Problem 10.2. While it perhaps is not obvious, $\hat{o}$ (10.37) is actually a weighted average of the g individual odds ratios,

$$o_i = \frac{p_{i1}(1 - p_{i2})}{p_{i2}(1 - p_{i1})}, \qquad i = 1, \ldots, g.$$

Show that this is so by finding a set of weights, $w_1, \ldots, w_g$, so that, with $\hat{o}$ given by (10.37),

$$\hat{o} = \frac{\sum\limits_{i=1}^{g} o_i w_i}{\sum\limits_{i=1}^{g} w_i}.$$

Problem 10.3. Apply the Mantel-Haenszel method to the data of Tables 10.6 and 10.7:

(a) What is the value of the odds ratio in each table? What should the summary value logically be? What is the value of $\hat{o}$ (10.37)?

(b) Since the sample sizes are large, Cochran's chi square analysis may be applied instead of Mantel's and Haenszel's. What are the values of d_1 and d_2 (10.14)? What are the values of w_1 and w_2 (10.16)? What are the values of χ^2_1 and χ^2_2 (10.17)? What is the value of $\chi^2_{\text{total}} = \chi^2_1 + \chi^2_2$? What is the value of

$$\chi^2_{\text{assoc}} = \frac{(m_1 w_1 + m_2 w_2)^2}{w_1 + w_2}?$$

What, finally, is the value of

$$\chi^2_{\text{homog}} = \chi^2_{\text{total}} - \chi^2_{\text{assoc}}?$$

What would you then conclude about the degree of association in Tables 10.6 and 10.7?

REFERENCES

Bartlett, M. S. (1935). Contingency table interactions. *J. roy. statist. Soc. Suppl.*, **2**, 248–252.

Berkson, J. (1968). Application of minimum logit χ^2 estimate to a problem of Grizzle with a notation on the problem of "no interaction." *Biometrics*, **24**, 75–95.

Berry, G. (1970). Parametric analysis of disease incidences in multiway tables. *Biometrics*, **26**, 572–579.

Bhapkar, V. P. and Koch, G. G. (1968a). Hypotheses of "no interaction" in multidimensional contingency tables. *Technometrics*, **10**, 107–122.

Bhapkar, V. P. and Koch, G. G. (1968b). On the hypotheses of "no interaction" in contingency tables. *Biometrics*, **24**, 567–594.

Birch, M. W. (1963). Maximum likelihood in three-way contingency tables. *J. roy. statist. Soc.*, *Ser. B*, **25**, 220–233.

Bishop, Y. M. M. (1969). Full contingency tables, logits, and split contingency tables. *Biometrics*, **25**, 383–399.

Cochran, W. G. (1954). Some methods of strengthening the common χ^2 tests. *Biometrics*, **10**, 417–451.

Cochran, W. G. (1968). The effectiveness of adjustment by subclassification in removing bias in observational studies. *Biometrics*, **24**, 295–313.

Cooper, J. E., Kendell, R. E., Gurland, B. J., Sharpe, L., Copeland, J. R. M. and Simon, R. (1972). *Psychiatric diagnosis in New York and London*. London: Oxford University Press.

Cox, D. R. (1958). The regression analysis of binary sequences. *J. roy. statist. Soc.*, *Ser. B*, **20**, 215–242.

Cox, D. R. (1970). *The analysis of binary data*. London: Methuen.

Dyke, G. V. and Patterson, H. D. (1952). Analysis of factorial arrangements when the data are proportions. *Biometrics*, **8**, 1–12.

Finney, D. J. (1965). The design and logic of a monitor of drug use. *J. chronic Dis.*, **18**, 77–98.

Fleiss, J. L. (1970). On the asserted invariance of the odds ratio. *Brit. J. prev. soc. Med.*, **24**, 45–46.

Gart, J. J. (1962). On the combination of relative risks. *Biometrics*, **18**, 601–610.

Gart, J. J. (1970). Point and interval estimation of the common odds ratio in the combination of 2 × 2 tables with fixed marginals. *Biometrika*, **57**, 471–475.

Gart, J. J. (1971). The comparison of proportions: A review of significance tests, confidence intervals and adjustments for stratification. *Rev. internat. statist. Inst.*, **39**, 16–37.

Goodman, L. A. (1963). On methods for comparing contingency tables. *J. roy. statist. Soc.*, *Ser. A*, **126**, 94–108.

Goodman, L. A. (1964). Simple methods for analyzing three-factor interaction in contingency tables. *J. Amer. statist. Assoc.*, **59**, 319–352.

Goodman, L. A. (1971). The analysis of multidimensional contingency tables: Stepwise procedures and direct estimation methods for building models for multiple classifications. *Technometrics*, **13**, 33–61.

Grizzle, J. E. (1961). A new method of testing hypotheses and estimating parameters for the logistic model. *Biometrics*, **17**, 372–385.

Grizzle, J. E. (1963). Tests of linear hypotheses when the data are proportions. *Amer. J. public Health*, **53**, 970–976.

Kastenbaum, M. A. and Lamphiear, D. E. (1959). Calculation of chi-square to test the no three-factor interaction hypothesis. *Biometrics*, **15**, 107–115.

Lewis, J. A. (1968). A program to fit constants to multiway tables of quantitative and quantal data. *Appl. Statist.*, **17**, 33–41.

Mantel, N. (1963). Chi-square tests with one degree of freedom: Extensions of the Mantel-Haenszel procedure. *J. Amer. statist. Assoc.*, **58**, 690–700.

Mantel, N. (1966). Models for complex contingency tables and polychotomous dosage response curves. *Biometrics*, **22**, 83–95.

Mantel, N. and Haenszel, W. (1959). Statistical aspects of the analysis of data from retrospective studies of disease. *J. natl. Cancer Inst.*, **22**, 719–748.

Maxwell, A. E. (1961). *Analysing qualitative data.* London: Methuen.

Maxwell, A. E. and Everitt, B. S. (1970). The analysis of categorical data using a transformation. *Brit. J. math. statist. Psychol.*, **23**, 177–187.

Naylor, A. F. (1967). Small sample considerations in combining 2×2 tables. *Biometrics*, **23**, 349–356.

Norton, H. W. (1945). Calculation of chi-square for complex contingency tables. *J. Amer. statist. Assoc.*, **40**, 251–258.

Pasternack, B. S. and Mantel, N. (1966). A deficiency in the summation of chi procedure. *Biometrics*, **22**, 407–409.

Radhakrishna, S. (1965). Combination of results from several 2×2 contingency tables. *Biometrics*, **21**, 86–98.

Rubin, D. B. (1970). *The use of matched sampling and regression adjustment in observational studies.* Ph.D. dissertation, Harvard University.

Sheehe, P. R. (1966). Combination of log relative risk in retrospective studies of disease. *Amer. J. public Health*, **56**, 1745–1750.

Tallis, G. M. (1964). The use of models in the analysis of some classes of contingency tables. *Biometrics*, **20**, 832–839.

Winsor, C. P. (1948). Factorial analysis of a multiple dichotomy. *Hum. Biol.*, **20**, 195–204.

Yates, F. (1955). The use of transformations and maximum likelihood in the analysis of quantal experiments involving two treatments. *Biometrika*, **42**, 382–403.

CHAPTER 11

The Effects of Misclassification Errors

We have so far assumed that the assignment of a subject to the diseased or nondiseased category, and to the category for the presence or absence of the antecedent factor, is made without error. This assumption is assuredly not valid. Whenever such assignments are made by the response to a questionnaire, or by responses in an interview, or by an examination of a case record, or by a physical or chemical test, or by any means imaginable, the assignment can be in error. For unintentional reasons—chance misreadings, failure to hear a response, and so on—or because of unconscious biases, a subject having the disease may be recorded as not having it, and conversely. Exactly the same is true for the recording of the antecedent factor.

In sampling methods I and II, misclassification errors can operate on either of the two characteristics studied. In sampling method III, they can operate only on the response variable.

In this chapter we consider the effects of misclassification errors, and in the next give some methods for estimating the extent of and for reducing error. Section 11.1 presents in some detail one example of the effects of misclassification. Section 11.2 describes algebraically the effects of misclassification errors on measures of association.

11.1. AN EXAMPLE OF THE EFFECTS OF MISCLASSIFICATION

Dern, Glynn, and Brewer (1963) studied the frequency of glucose-6-phosphate dehydrogenase (G-6-PD) deficiency in the erythrocytes of male Negro schizophrenic patients in the Chicago area. G-6-PD deficiency is inherited as a sex-linked error of metabolism, and is found in about 10% to 15% of the American male Negro population.

The deficiency is sometimes referred to as the fava bean disease because exposure to the fava bean by affected individuals tends to result in hemolysis, that is, in a breakdown of red blood cells. Antimalarial agents and other drugs also result in hemolysis in affected individuals.

The data provided by Dern, Glynn, and Brewer are summarized in Table 11.1. For these data, chi square = 7.95, which indicates an association significant at the .01 level.

Table 11.1. *Association between G-6-PD deficiency and subtype of schizophrenia in Chicago*

	Diagnosis		
G-6-PD Status	Catatonic	Paranoid	Total
Deficient	15	6	21
Nondeficient	57	99	156
Total	72	105	177

The contrasted proportions are the proportion of the catatonics who are deficient, $p_C = 15/72 = .208$, and the proportion of the paranoids who are deficient, $p_P = 6/105 = .057$. The value of the odds ratio is

$$o = \frac{15 \times 99}{6 \times 57} = 4.34; \tag{11.1}$$

that is, the odds that a catatonic is G-6-PD deficient are over four times the odds that a paranoid is deficient.

Fieve, Brauninger, Fleiss, and Cohen (1965) repeated the study at five state hospitals in the New York City area. At four of the hospitals, the results were as given in Table 11.2.

Table 11.2. *Association between G-6-PD deficiency and subtype of schizophrenia in four New York state hospitals*

Hospital	Catatonic		Paranoid		
	N	% Deficient	N	% Deficient	o
Central Islip	32	15.6	80	12.5	1.30
Pilgrim	78	16.7	76	6.6	2.84
Brooklyn	13	30.8	18	11.1	3.56
Kings Park	55	10.9	96	6.3	1.84

The four individual odds ratios did not differ significantly; the mean odds ratio,

$$\bar{o} = 2.09, \tag{11.2}$$

was significantly different from unity at the .05 level (see Chapter 10 for methods of comparing and combining different odds ratios). The findings from these four hospitals tend to confirm Dern, Glynn, and Brewer's (1963) original finding, but indicate a reduced degree of association.

At the fifth hospital, Rockland, the odds ratio was again significantly different from unity (Table 11.3). The problem at Rockland State Hospital,

Table 11.3. *Association between G-6-PD deficiency and subtype of schizophrenia at Rockland State Hospital*

Catatonic		Paranoid		
N	% Deficient	N	% Deficient	o
28	7.1	29	24.1	0.24

clearly, was that the odds ratio was significantly different from unity in the reverse direction from that found previously.

The investigators quickly checked back with the administration at Rockland, and breathed a sigh of relief when they discovered that half of the schizophrenic patients had been withheld because they were subjects in other research investigations. The investigators therefore returned to Rockland and succeeded in studying all of the resident male Negro catatonics and paranoids. The results from the second survey were as shown in Table 11.4.

Table 11.4. *Association between G-6-PD deficiency and subtype of schizophrenia at Rockland State Hospital—second survey*

Catatonic		Paranoid		
N	% Deficient	N	% Deficient	o
37	2.7	87	16.1	0.14

The odds ratio was again significantly different from unity at the .05 level, but even further below it than before.

Thus there is evidence for an association between an inherited enzyme deficiency and subtype of schizophrenia, but, unfortunately, for association in both possible directions. What are missing to complete a confusing picture are data from a large sample of patients showing no significant difference between paranoids and catatonics. Just such data were provided by a sample of 426 patients at a Veterans Administration hospital in Alabama (Bowman et al., 1965). The data are presented in Table 11.5. For these data, $p_C = 10.4\%$, $p_P = 11.8\%$, $o = 0.87$, and chi square $= 0.07$.

Table 11.5. *Association between G-6-PD deficiency and subtype of schizophrenia at a VA hospital in Alabama*

	Diagnosis		
G-6-PD Status	Catatonic	Paranoid	Total
Deficient	17	31	48
Nondeficient	146	232	378
Total	163	263	426

There is, therefore, evidence in the literature for all three kinds of association: positive, negative, and zero. This conflicting evidence is summarized in Table 11.6.

Table 11.6. *Evidence for various directions of association between G-6-PD deficiency and subtype of schizophrenia*

Direction of Association	Source	o
Catatonics over paranoids	Chicago	4.34
	Four New York hospitals	2.09
Paranoids over catatonics	Fifth New York hospital	0.14
No difference	Alabama	0.87

In attempting to account for this confusion, the New York investigators looked first at the experimental techniques used in the three studies. The techniques, while different, were not sufficiently dissimilar to explain the discrepancies. At any rate, the technique used at Rockland was no different from that used at the other four hospitals in New York.

Differences in the drugs given the patients might conceivably have produced these discrepant findings. Within each study, therefore, odds ratios were calculated for each major category of drug administered at the time of the study. With only a few exceptions, on too few cases to have much impact, the odds ratios for the specific drug categories within a study were in the same direction as the overall value for the study.

Whatever the influences might be of differences in techniques of blood testing and of drugs, they pale when the unreliability of psychiatric diagnosis is considered. A large literature has accumulated indicating just how unreliable psychiatric diagnosis is (Zubin, 1967). With respect to schizophrenia, for example, it has been found that, typically, only about 70% of all patients given a diagnosis of schizophrenia by one diagnostician are given the same diagnosis by a second, and that perhaps 10% of patients given a diagnosis other than schizophrenia by one diagnostician are given a diagnosis of schizophrenia by a second. From the few published data for the subtypes of schizophrenia, it appears that reliability is less for paranoid and catatonic schizophrenia than for schizophrenia in general.

In each of the three studies cited of G-6-PD deficiency and schizophrenia, the then current hospital diagnoses were accepted uncritically, with no attempt made to verify their accuracy. It is thus likely that the single most important source of the discrepancies among the three studies, as well as between Rockland and the other four New York state hospitals, was the unreliability of psychiatric diagnosis.

Prima facie evidence for diagnostic differences among the five New York hospitals is afforded by the variability of the proportions of patients diagnosed catatonic among those diagnosed either catatonic or paranoid. There exist some differences among the five hospitals in the kinds of patients they receive, but not of sufficient magnitude to account for the differences in their proportions of catatonics. Problem 11.1 is devoted to these differences.

The example just given is from psychiatry, but the inference should not be drawn that psychiatry is unique in being plagued by inaccurate diagnoses. Unreliability exists in the diagnosis of childhood disorders (Derryberry, 1938; Bakwin, 1945); in the diagnosis of emphysema (Fletcher, 1952); in the interpretation of electrocardiograms (Davies, 1958); in the interpretation of X-rays (Yerushalmy, 1947; Cochrane and Garland, 1952); and in the certification of causes of death (Markush, Schaaf, and Seigel, 1967). A review of diagnostic unreliability in other branches of physical medicine is given by Garland (1960).

It may, in fact, be taken as axiomatic that the determination of the presence or absence of any disease or condition and the determination of the exact form of the condition when it is present are subject to error. Likewise, the determination of the presence or absence of an antecedent factor, with the possible exception of a subject's sex, is subject to error.

11.2. THE ALGEBRA OF MISCLASSIFICATION

Confusion exists about the effects of errors of misclassification. The facts of the matter are that such errors can turn a truly strong positive association into one that is less strongly positive or even apparently negative; one that is truly strongly negative into one that is less strongly negative or even apparently positive; and one that is nil into one that is apparently strong. These facts contradict the long-standing, but erroneous impression that errors of misclassification tend only to reduce the magnitude of association (Newell, 1962).

Assume for simplicity that the classification of a person as diseased or not is accurate, so that the only source of error is in the determination of whether or not the factor under study was present. To keep matters specific, let us consider the comparison of women aged 55–64 who were newly diagnosed as having lung cancer with similarly aged women newly diagnosed as having cancer of the breast, with respect to whether they ever smoked.

We assume that the diagnoses are accurate, but that the determination of smoking history is subject to error. Two sources of error exist, one residing with the informant and one with the person taking the history. With respect to the informant (e.g., the patient herself or a relative):

1. He may misunderstand the intent or phrasing of the question.
2. He may make an honest mistake in reporting what the patient's smoking status was.
3. He may deliberately misrepresent the patient's smoking status (more likely in claiming the patient never smoked when in fact she did than the reverse).

With respect to the person taking the history:

1. He may misunderstand the informant's answer.
2. He may make an honest coding error.
3. He may apply different standards in the recording of certain responses for one type of patient versus another. Suppose, for example, that the history-taker elects to bend over backwards in an attempt to control for a possible bias toward finding an association. He may then record the statement, "I smoked once in a while when I was a kid, but never since," as Never Smoked if made by a lung cancer patient but as Ever Smoked if made by a patient with breast cancer.

Methods for reducing the effects of some of these sources of error are discussed in the next chapter. Here we study just what these effects are. The analysis is that of Keys and Kihlberg (1963). Consider first the lung cancer patients, who, we are assuming, can be identified without error. Let P_L denote the true proportion of lung cancer patients who had ever smoked, so that $1 - P_L$ is the true proportion who had never smoked.

Denote by E_L the false negative rate and by F_L the false positive rate for the lung cancer patients; that is, E_L is the probability that a lung cancer patient who actually had smoked is recorded as not having smoked, and F_L is the probability that a lung cancer patient who actually had never smoked is recorded as having smoked. Whereas one would wish to estimate P_L, the true proportion who had ever smoked, one can instead, from the recorded histories, only estimate

$$p_L = (1 - E_L)P_L + F_L(1 - P_L).\tag{11.3}$$

The estimated proportion of lung cancer patients who had ever smoked, p_L, is a fraction, $1 - E_L$, of those who truly ever smoked plus a fraction, F_L, of those who truly never smoked.

The observable proportion p_L may be less than, equal to, or greater than the true proportion P_L depending on the relative magnitudes of E_L and F_L. In fact,

$$p_L > P_L \quad \text{if} \quad \frac{F_L}{E_L + F_L} > P_L,$$

$$p_L = P_L \quad \text{if} \quad \frac{F_L}{E_L + F_L} = P_L,$$

and

$$p_L < P_L \quad \text{if} \quad \frac{F_L}{E_L + F_L} < P_L.$$

If E_L and F_L are of approximately the same magnitude, there will be overestimation if P_L is less than .5 and underestimation if P_L is greater than .5. Thus, even if the error rates are equal, the errors will not necessarily cancel out.

Let, now, P_B denote the true proportion of breast cancer patients who had ever smoked, E_B their false negative rate, and F_B their false positive rate. For the breast cancer patients, therefore, one can estimate only the proportion recorded as ever having smoked, say

$$p_B = (1 - E_B)P_B + F_B(1 - P_B).\tag{11.4}$$

The algebra of the effects of errors on the odds ratio is complicated (Diamond and Lilienfeld, 1962a, 1962b). Suppose, therefore, that the association between smoking and type of cancer is measured simply by the difference between the proportions who had smoked. Instead of being able to estimate the true difference, say

$$D = P_L - P_B,\tag{11.5}$$

we can only estimate $d = p_L - p_B$. This difference between the recorded proportions is easily seen to reduce algebraically to

$$d = D + (F_L - F_B) + P_B(E_B + F_B) - P_L(E_L + F_L),\tag{11.6}$$

which indicates that d is typically a biased estimate of D.

The estimated difference d may be less than, equal to, or greater than the true difference D. It may even be of opposite sign, which means that an association that is actually in one direction may be estimated as being in the opposite direction.

This possibility of a reversal of the direction of association cannot arise in the special case in which the two false positive rates and the two false negative rates are equal. Suppose that the two false negative rates are equal,

$$E_L = E_B = E, \tag{11.7}$$

say, and that the two false positive rates are equal,

$$F_L = F_B = F, \tag{11.8}$$

say. By substituting (11.7) and (11.8) into (11.6) and simplifying, we find that the difference between the recorded proportions is

$$d = D(1 - (E + F)). \tag{11.9}$$

The first point to notice about (11.9) is that d, the difference that can be calculated, cannot possibly equal D, the true difference, whenever either error rate is nonzero. The second point to notice is that, provided both E and F are less than $\frac{1}{2}$—that is, both error rates are less than 50%—the observed difference is in the same direction as the true difference, but is numerically smaller—that is, is closer to zero. This is the situation considered by Bross in a classic paper (1954) and the one which has led to the erroneous anticipation that misclassification errors *always* tend to deflate the difference between two rates. Equal false positive rates and equal false negative rates must, however, be considered unusual (see, e.g., Lilienfeld and Graham, 1958).

With respect to the odds ratio, the effect again can in general be anything. In the particular case just considered, where the false positive rates are equal and less than $\frac{1}{2}$, and where the false negative rates are equal and less than $\frac{1}{2}$, the same kind of underestimation as for the difference between rates occurs. Specifically, if ω is the true odds ratio and o the odds ratio estimated from misclassified data, then, if $\omega > 1$, we would expect to find $\omega > o > 1$. That is, the estimated odds ratio would also be greater than unity, but not by as much as the true value.

Further analyses of the effects of misclassification on measures of association and on chi square tests have been performed by Rogot (1961), Mote and Anderson (1965), Assakul and Proctor (1967), and Koch (1969). This discussion has been concerned only with errors in one of the two variables under study. A discussion of the more realistic situation in which both variables are subject to errors of misclassification is given by Keys and Kihlberg (1963).

Problem 11.1. The text cited the differences among the five New York state hospitals in the proportions of patients diagnosed catatonic, out of all those diagnosed either catatonic or paranoid.

(*a*) The frequencies are given below. Calculate the indicated proportions.

Hospital	Catatonic	Paranoid	Total	Proportion Catatonic
Central Islip	32	80	112	$= p_1$
Pilgrim	78	76	154	$= p_2$
Brooklyn	13	18	31	$= p_3$
Kings Park	55	96	151	$= p_4$
Rockland	37	87	124	$= p_5$
Total	215	357	572	$= \bar{p}$

(*b*) The chi square statistic for comparing a series of proportions is given by (9.4). Calculate chi square for the proportions determined in (*a*).

(*c*) Refer the value just calculated to Table A.1 with four degrees of freedom. At what significance level would the hypothesis of no difference in proportions be rejected? What would you conclude about the standards for the differential diagnosis of catatonic and paranoid schizophrenia in the five hospitals?

Problem 11.2. Suppose that the rate of smoking among women aged 55–64 who were newly diagnosed as having lung cancer is $P_L = .50$, and suppose that the error rates are $E_L = .25$ and $F_L = .05$.

(*a*) What is the value of p_L (11.3), the estimated proportion of such women who ever smoked?

Suppose that the rate of smoking among women aged 55–64 who were newly diagnosed as having breast cancer is $P_B = .40$, and suppose that the error rates are $E_B = F_B = .10$.

(*b*) What is the value of p_B (11.4), the estimated proportion of such women who ever smoked?

(*c*) What is the actual difference between the rates? What is the estimated difference? How do these compare?

(*d*) What is the actual odds ratio? What is the estimated odds ratio? How do these compare?

REFERENCES

Assakul, K. and Proctor, C. H. (1967). Testing independence in two-way contingency tables with data subject to misclassification. *Psychometrika*, **32**, 67–76.

Bakwin, H. (1945). Pseudodoxia pediatrica. *New Engl. J. Med.*, **232**, 691–697.

Bowman, J. E., Brewer, G. J., Frischer, H., Carter, J. L., Eisenstein, R. B. and Bayrakci, C. (1965). A re-evaluation of the relationship between glucose-6-phosphate dehydrogenase deficiency and the behavioral manifestations of schizophrenia. *J. Lab. clin. Med.*, **65**, 222–227.

Bross, I. (1954). Misclassification in 2 × 2 tables. *Biometrics*, **10**, 478–486.

Cochrane, A. L. and Garland, L. H. (1952). Observer error in interpretation of chest films: International investigation. *Lancet*, **2**, 505–509.

Davies, L. G. (1958). Observer variation in reports on electrocardiograms. *Brit. Heart J.*, **20**, 153–161.

Dern, R. J., Glynn, M. F. and Brewer, G. J. (1963). Studies on the correlation of the genetically determined trait G-6-PD deficiency with behavioral manifestations in schizophrenia. *J. Lab. clin. Med.*, **62**, 319–329.

Derryberry, M. (1938). Reliability of medical judgments on malnutrition. *Public Health Rep.*, **53**, 263–268.

Diamond, E. L. and Lilienfeld, A. M. (1962a). Effects of errors in classification and diagnosis in various types of epidemiological studies. *Amer. J. public Health*, **52**, 1137–1144.

Diamond, E. L. and Lilienfeld, A. M. (1962b). Misclassification errors in 2 × 2 tables with one margin fixed: Some further comments. *Amer. J. public Health*, **52**, 2106–2110.

Fieve, R. R., Brauninger, G., Fleiss, J. L. and Cohen, G. (1965). Glucose-6-phosphate dehydrogenase deficiency and schizophrenic behavior. *J. psychiatr. Res.*, **3**, 255–262.

Fletcher, C. M. (1952). Clinical diagnosis of pulmonary emphysema—an experimental study. *Proc. roy. Soc. Med.*, **45**, 577–584.

Garland, L. H. (1960). The problem of observer error. *Bull. N.Y. Acad. Med.*, **36**, 570–584.

Keys, A. and Kihlberg, J. K. (1963). The effect of misclassification on estimated relative prevalence of a characteristic. *Amer. J. public Health*, **53**, 1656–1665.

Koch, G. G. (1969). The effect of non-sampling errors on measures of association in 2 × 2 contingency tables. *J. Amer. statist. Assoc.*, **64**, 852–863.

Lilienfeld, A. M. and Graham, S. (1958). Validity of determining circumcision status by questionnaire as related to epidemiological studies of cancer of the cervix. *J. natl. Cancer Inst.*, **21**, 713–720.

Markush, R. E., Schaaf, W. E. and Seigel, D. G. (1967). The influence of the death certifier on the results of epidemiologic studies. *J. natl. med. Assoc.*, **59**, 105–113.

Mote, V. L. and Anderson, R. L. (1965). An investigation of the effect of misclassification on the properties of chi square tests in the analysis of categorical data. *Biometrika*, **52**, 95–109.

Newell, D. J. (1962). Errors in the interpretation of errors in epidemiology. *Amer. J. public Health*, **52**, 1925–1928.

Rogot, E. (1961). A note on measurement errors and detecting real differences. *J. Amer. statist. Assoc.*, **56**, 314–319.

Yerushalmy, J. (1947). Statistical problems in assessing methods of medical diagnosis, with special reference to X-ray techniques. *Public Health Rep.*, **62**, 1432–1449.

Zubin, J. (1967). Classification of the behavior disorders. Pp. 373–406 in P. R. Farnsworth, O. McNemar and Q. McNemar (Eds.). *Annual review of psychology*. Palo Alto, Calif.: Annual Reviews.

CHAPTER 12

The Measurement and Control of Misclassification Error

In the preceding chapter we considered some possible effects of misclassification errors. In Section 12.1 we consider statistical means of controlling for error. Some methods for estimating the magnitude of error are presented in Section 12.2, and some techniques for the experimental control of error are discussed in Section 12.3.

12.1. STATISTICAL CONTROL FOR ERROR

Occasionally, an investigator has available two or more means for determining the status of a patient—one quite expensive but reliable (i.e., subject to little error), the others relatively inexpensive but unreliable. If he is planning a survey or comparative study on even a moderate scale, the investigator must, to keep the cost of the study as low as possible, employ one of the unreliable devices (Rubin, Rosenbaum, and Cobb, 1956).

If the investigator uses *only* an unreliable device, he runs the risk of obtaining the kinds of biased estimates described in the preceding chapter. By selecting a subsample of the total sample to be assessed by both the unreliable and more reliable devices, however, he will be able to estimate, for only a relatively small added cost, the rates of misclassification and thus to correct for bias.

Consider, as an example, the determination of the current smoking habits of each of a sample of subjects. The investigator can choose to rely solely on the subject's report, but would be sacrificing reliability for simplicity. He can instead rely on a chemical test for the concentrations of thiocyanates in the subject's urine, saliva, and plasma (Densen, Davidow, Bass, and Jones, 1967), but would be sacrificing inexpensiveness for precision.

Suppose that a sample of N newly hospitalized women diagnosed as having lung cancer is to be evaluated for smoking habits, and suppose that the investigator chooses to rely on each woman's verbal report on her current smoking practice. For simplicity, he characterizes each woman as either a heavy smoker (say, smoking ten or more cigarettes per day, on the average) or not. Let p_L denote the resulting proportion who report being heavy smokers.

The selected means of determining smoking status has the virtue of being inexpensive, but the drawback of being prone to possibly excessive error. Suppose, therefore, that the investigator decides to estimate the degree of error present in the patients' reports by taking a subsample of n out of the total of N lung cancer patients and testing, in addition, their plasma concentrations of thiocyanates. A positive result on the test is taken as indicative of the patient's being a heavy smoker, and a negative result as indicative of her not being a heavy smoker.

The results of this blood test can hardly be viewed as establishing the patient's true status, not only because the dichotomy between heavy and non-heavy smoking is imprecise but also because the results of the test are subject to random fluctuations themselves. Nevertheless, because of its greater reproducibility, the blood test may be viewed as a standard against which to compare verbal reports.

Let Table 12.1 represent the cross-classification of reported and tested smoking status for the subsample of n lung cancer patients. The notation

Table 12.1. Smoking status as determined by report and by the standard blood test

Report	Standard Heavy	Standard Not Heavy	Total
Heavy	n_{00}	n_{01}	n_0
Not heavy	n_{10}	n_{11}	n_1

is that of Tenenbein (1970, 1971). From these data we may estimate as n_{00}/n_0 the proportion, of those women whose reported status is that of heavy smoker, who would be assigned to the heavy smoking category by the standard, and as n_{10}/n_1 the proportion, of those women whose reported status is that of nonheavy smoker, who would be assigned to the heavy smoking category by the standard. Recalling that p_L is the overall proportion of women assigned to the heavy smoking category on the basis of verbal report, it is easily checked that an estimate of the overall proportion who would have been so assigned by the standard is

$$P_L = \frac{n_{00}}{n_0} p_L + \frac{n_{10}}{n_1} (1 - p_L). \tag{12.1}$$

Whereas the estimated standard error of p_L is simply $\sqrt{p_L(1 - p_L)/N}$, that of P_L is more complicated. It is, in fact, given by

$$\text{s.e.}(P_L) = \sqrt{\frac{P_L(1 - P_L)}{N} \left(1 + (1 - K) \frac{N - n}{n}\right)}, \tag{12.2}$$

where

$$K = \frac{1 - p_L}{p_L} \frac{(P_L - n_{10}/n_1)^2}{P_L(1 - P_L)}. \tag{12.3}$$

The estimate (12.1) and standard error (12.2) are derived by Tenenbein (1970, 1971), who also gives criteria for choosing a reasonable value of n. Different approaches to the estimation of correction factors are described by Harper (1964), Bryson (1965), and Press (1968).

The following numerical example will illustrate the algebra presented above. Suppose that a total of 200 female lung cancer patients are interviewed, and that 88 of them so respond that they are judged to be heavy smokers. The observed, but biased rate of heavy smoking among lung cancer patients is therefore

$$p_L = .44. \tag{12.4}$$

Suppose, however, that 50 of the 200 patients are tested for levels of serum thiocyanates as well as interviewed, and suppose that the resulting cross-classification is as given in Table 12.2. Then, $n_{00}/n_0 = 18/20 = .90$ and

Table 12.2. **Smoking status of 50 lung cancer patients as determined by report and by chemical test**

	Chemical Test		
Report	Heavy	Not Heavy	Total
Heavy	18	2	20
Not heavy	6	24	30

$n_{10}/n_1 = 6/30 = .20$ are the two correction factors, and substitution into (12.1) yields

$$P_L = .90 \times .44 + .20 \times .56 = .51 \tag{12.5}$$

as an improved estimate of the rate of heavy smoking in this group. Note that the rate given in (12.4) is an underestimate by more than 10%.

To determine the standard error of P_L, the quantity K given by (12.3) must first be calculated. It is

$$K = \frac{.56}{.44} \frac{(.51 - .20)^2}{.51 \times .49} = .4894.$$ (12.6)

Substitution of this value in (12.2) yields

$$\text{s.e.}(P_L) = \sqrt{\frac{.51 \times .49}{200} \left(1 + .5106 \times \frac{150}{50}\right)}$$

$$= \sqrt{.0032} = .06$$ (12.7)

as the estimated standard error of P_L.

If the study is a comparative one, such as comparing the rates of heavy smoking among lung cancer and breast cancer patients, then a subsample of the breast cancer patients would also have to be administered the blood test. Problem 12.1 gives some comparative data for analysis.

12.2. ESTIMATING THE MAGNITUDE OF ERROR

The method of the preceding section is applicable only when a well defined standard means of characterizing a subject is available for use on a subsample of a total group. Such a standard may be a chemical test, a physical examination, or an autopsy.

For some variables of importance, however, no such standard is readily apparent. In a retrospective study of the genetics of manic-depressive disorders, for example, an investigator often has no alternative but to rely on a subject's or his siblings' report as to whether his parents were ever pathologically excited or depressed. In a survey aimed at estimating the rate of impaired efficiency at work among workers of a specified ethnic background, the investigator must rely on the reports of the subject, his coworkers and employer, and perhaps on direct observation for a short amount of time at the place of employment, and must then assimilate these reports and observations into a summary, often subjective rating of impairment.

To assess the extent to which a given rater's characterization of a subject is reliable, it is clear that we must have a number of subjects characterized by more than one rater. The degree of agreement among the raters provides no more than an upper bound to the degree of accuracy present in the ratings, however. If agreement among the raters is high, then there is a possibility, but by no means a guarantee, that the ratings do in fact reflect the dimension they are purported to. If their agreement is low, on the other hand, then the usefulness of the ratings is severely limited, for it is meaningless to ask what is associated with the variable being rated when one cannot even trust those ratings to begin with.

Suppose for simplicity that each of a sample of individuals is rated by two raters, with each subject being assigned to one of three mutually exclusive categories by both raters. Consider the hypothetical example given in Table 12.3.

Table 12.3. Ratings on 100 subjects by two raters

	Rater A			
Rater B	1	2	3	Total
1	20	12	8	40
2	11	6	13	30
3	19	2	9	30
Total	50	20	30	100

Such evidence for the agreement between two raters is only infrequently obtained. Even when obtained, such evidence is only infrequently analyzed correctly. What one usually finds, in fact, is that reliability is measured in terms of the magnitude of chi square. What makes chi square inadequate is that it measures association of any kind, and not specifically agreement.

The first step in the calculation of chi square is the determination of the frequencies expected in each of the cells under the hypothesis of no association. The frequency expected in the ith row and jth column is calculated as

$$n_{ij}(e) = \frac{n_{i.} n_{.j}}{n_{..}},$$

where $n_{i.}$ is the total frequency in the ith row, $n_{.j}$ is the total frequency in the jth column, and $n_{..}$ is the overall frequency. An application of this formula to the frequencies of Table 12.3 yields the expected frequencies given in Table 12.4. For example, $n_{12}(e) = 40 \times 20/100 = 8$.

Table 12.4. Expected frequencies for data of Table 12.3

	Rater A			
Rater B	1	2	3	Total
1	20	8	12	40
2	15	6	9	30
3	15	6	9	30
Total	50	20	30	100

Next, each observed frequency, say n_{ij}, is compared with its corresponding expected frequency, $n_{ij}(e)$, by means of the expression

$$d_{ij} = \frac{(n_{ij} - n_{ij}(e))^2}{n_{ij}(e)}. \tag{12.8}$$

Thus $d_{11} = (20 - 20)^2/20 = 0$, $d_{12} = (12 - 8)^2/8 = 2$, and so on.

Finally, all values of d_{ij} are summed to yield a value of chi square with, in general, $(m - 1)^2$ degrees of freedom, where m is the number of categories rated:

$$\chi^2 = 0 + 2 + \cdots + 2.67 + 0 = 9.91.$$

The obtained value of chi square, with $(3 - 1)^2 = 2^2 = 4$ degrees of freedom, is significant at the .05 level, indicating that an association exists between the assignments made by the two raters. Chi square tells us nothing at all, however, about their degree of agreement.

In order to measure agreement, we must focus attention on the cells along the diagonal, representing assignments agreed upon by the two raters. In doing so we notice that, for each of the three categories, the number of individuals on whom there is agreement (20, 6, and 9 for the three respective categories, from Table 12.3) is precisely equal to the number on whom agreement is to be expected solely on the basis of chance (again 20, 6, and 9 from Table 12.4). We would therefore have to conclude that the degree of agreement between the two raters is no better than that predicted by chance, or random ratings. Clearly, reliance on the significance of chi square would have led to the totally erroneous conclusion that the ratings were reliable.

Just as chi square is inadequate for measuring reliability, so is the overall proportion of individuals on whom there is agreement (Cohen, 1960; Armitage, Blendis, and Smyllie, 1966; Rogot and Goldberg, 1966). The proportion of agreement is, say,

$$p_o = \frac{n_{11} + \cdots + n_{mm}}{n_{..}}, \tag{12.9}$$

where m represents the total number of categories. For the data of Table 12.3,

$$p_o = \frac{20 + 6 + 9}{100} = .35. \tag{12.10}$$

The inadequacy of p_o as a measure of reliability is seen in the data of Table 12.4: some degree of agreement is expected purely on the basis of chance. In fact, the overall proportion of agreement expected by chance alone is, say,

$$p_c = \frac{n_{1.}n_{.1} + \cdots + n_{m.}n_{.m}}{n_{..}^2}. \tag{12.11}$$

In the example,

$$p_c = \frac{40 \times 50 + 30 \times 20 + 30 \times 30}{100 \times 100} = .35, \qquad (12.12)$$

precisely equal to the value for p_o in (12.10).

A better measure of agreement than p_o alone is $p_o - p_c$, that is, how much agreement exists beyond the amount expected by chance alone. In our example $p_o - p_c = 0$, so our conclusion would be that *all* the observed agreement could be attributed to chance alone. Since we are accustomed to indexing perfect agreement by the value $+1$, and since $p_o - p_c$ does not possess this property, the statistic *kappa* (see Cohen, 1960) is suggested for routine use:

$$\kappa = \frac{p_o - p_c}{1 - p_c}. \qquad (12.13)$$

If $p_o < p_c$, indicating even less than chance agreement, then $\kappa < 0$. If $p_o = p_c$, indicating just chance agreement, then $\kappa = 0$. If $p_o > p_c$, indicating more than chance agreement, then $\kappa > 0$. If, finally, $p_o = 1$, indicating perfect agreement, then $\kappa = 1$.

To assess the statistical significance of κ, we require its standard error. If we define

$$p_{i.} = \frac{n_{i.}}{n_{..}},$$

that is, the proportion of all subjects assigned to the ith category by rater B, and

$$p_{.i} = \frac{n_{.i}}{n_{..}},$$

that is, the proportion of all subjects assigned to the ith category by rater A, then a simple expression for p_c is, by (12.11),

$$p_c = \sum_{i=1}^{m} p_{i.} p_{.i}, \qquad (12.14)$$

and the standard error of κ is given by the square root of

$$\text{Var}(\kappa) = \frac{1}{N(1 - p_c)^2}\left(p_c + p_c^2 - \sum_{i=1}^{m} p_{i.} p_{.i}(p_{i.} + p_{.i})\right) \qquad (12.15)$$

(Everitt, 1968; Fleiss, Cohen, and Everitt, 1969). Applying this formula to the data of Table 12.3, we find that

$$\text{Var}(\kappa) = \frac{1}{100 \times .65^2}$$

$$\times (.35 + .35^2 - (.4 \times .5 \times .9 + .3 \times .2 \times .5 + .3 \times .3 \times .6))$$

$$= \frac{1}{42.25}(.35 + .1225 - (.180 + .030 + .054))$$

$$= .0049, \tag{12.16}$$

so that the standard error of κ is s.e.$(\kappa) = \sqrt{.0049} = .07$. The statistical significance of κ may be determined by referring the quantity

$$z = \frac{\kappa}{\text{s.e.}(\kappa)} \tag{12.17}$$

to Table A.2 of the standard normal curve. If z is significantly large, the conclusion would be that the observed degree of agreement reflects bona fide reliability. If z is small, the conclusion would be that the observed degree of agreement might only reflect random ratings.

Kappa has been generalized in a number of directions. When the relative seriousness of the different kinds of disagreements can be specified, the statistic *weighted kappa* (Spitzer, Cohen, Fleiss, and Endicott, 1967; Cohen, 1968) is appropriate. A second generalization (by Fleiss, Spitzer, Endicott, and Cohen, 1972) is appropriate when each subject is characterized by multiple diagnoses or by multiple causes of death (see, e.g., Guralnick, 1966). Kappa has also been generalized to the case where more than two raters rate each subject (Fleiss, 1971; Light, 1971).

When agreement on quantitative scales is assessed, the *intraclass correlation coefficient* is employed as a measure of reliability (Ebel, 1951; Haggard, 1958; Burdock, Fleiss, and Hardesty, 1963; and Bartko, 1966). Krippendorff (1970) has shown that kappa and weighted kappa are interpretable as intraclass correlation coefficients.

Cochran (1968) has made a general survey of measurement and classification errors, describing their effects on standard statistical techniques and suggesting some remedial procedures. Fleiss (1966) and Maxwell and Pilliner (1968) present methods specific to the case where agreement on a number of ratings is assessed, and Fleiss (1970) suggests a model appropriate to the case where the ratings are based on information obtained during an interview. Some methods for designing reliability studies are suggested by Fleiss (1963, 1970) and by Fleiss, Spitzer, and Burdock (1965).

12.3. THE EXPERIMENTAL CONTROL OF ERROR

It is almost always possible to modify a contemplated research design so as to reduce the probable magnitude of error. Only a few of the large number of techniques and ideas that exist can be presented here. One procedure is modeled on the double-blind feature of a properly designed clinical trial. In a double-blind trial, the patient is ignorant as to which of the drugs being compared he is getting, and the person conducting the evaluation of the patient's response is likewise ignorant of the patient's treatment.

This idea of keeping both the patient and the evaluator in the dark has obvious relevance in studies in which the presence or absence of a disease and the presence or absence of an antecedent factor have to be determined at nearly the same time (e.g., the patients being studied may be admissions to an acute treatment ward in a hospital, and neither previous records nor the opportunity for follow-up may exist). One danger is that the diagnostician, knowing whether or not the factor is present, may be prejudiced in favor of one or another of the diagnoses under study. A control is to instruct the diagnostician not to ask about the antecedent factor unless it happens to be pathognomonic.

A second danger is that the person seeking to establish the presence or absence of the factor, knowing that the diagnosis has been made, may be prejudiced in favor of recording the factor as present, and vice versa. A control is to keep this second person ignorant of the diagnosis.

A third danger is that the patient may respond differently depending on whether he knows, or even believes, that he has the disease being studied. A control is to keep the patient ignorant of his diagnosis long enough for all the background information of interest to be collected. An ethical problem must be solved here: just how much can be withheld from a patient, and for how long, has to be determined on an ad hoc basis (see Levin, 1954).

We have so far taken for granted that the person responsible for making the diagnosis is different from the person responsible for eliciting information on the background factors. The two roles are not always capable of separation, but the results of a study in which the same person assumed both roles must always be suspect. An example of the bias which may arise is provided by a study of the psychiatric concomitants of systemic lupus erythematosus (SLE).

There has been an increasing number of reports that the frequency (on both an incidence and prevalence basis) of psychological disturbance among SLE patients is unusually high. As part of a study of this phenomenon (Ganz, Gurland, Deming, and Fisher, 1972), a psychologist interviewed, using a structured interview schedule, samples of SLE and rheumatoid arthritis (RA) patients coming to clinics at four New York City hospitals. RA patients were selected as a control group because of many similarities between the two diseases, and because few reports exist of psychiatric complications in RA.

Sixty-eight SLE and 36 RA patients were interviewed, on the basis of which each patient's psychiatric symptomatology was characterized as severe (e.g., extreme anxiety or depression, and, occasionally, delusional thinking), moderate (e.g., slight degrees of worrying and depression), or none. Every attempt was made to keep the interviewer ignorant of the patients' diagnoses. Table 12.5 gives the results of the categorization of symptomatology elicited by the interview.

Table 12.5. Psychiatric symptomatology by diagnosis
—results from interview

Symptoms	Systemic Lupus Erythematosus		Rheumatoid Arthritis	
	N	(%)	*N*	(%)
Severe	24	(35%)	9	(25%)
Moderate	19	(28%)	12	(33%)
None	25	(37%)	15	(42%)
Total	68	(100%)	36	(100%)

The proportions for the two diagnostic categories are seen to be similar. The chi square statistic, with two degrees of freedom, for comparing the two distributions is equal to 1.16, which is not significant at any reasonable level. Thus, on the basis of a structured interview conducted by an interviewer ignorant of the diagnosis, the conclusion would have to be that the degree of psychiatric symptomatology in SLE patients is not essentially different from that in RA patients.

As part of the study, the notes made by the physician on the same day as the interview were also examined. The physician was not told the results of the interview. With the same criteria as used before for characterizing the degree of psychiatric symptomatology, only now applied to the case notes, the data of Table 12.6 resulted. The assessment of the information recorded on the case notes was made by a person other than the interviewer.

The proportions for the two diagnostic categories are seen to be quite different. For example, nearly a third of the SLE patients as opposed to only 6% of the RA patients were characterized on the basis of case notes as having psychiatric symptoms of any kind. The value of chi square for the data of Table 12.6 is 8.27, indicating a difference significant at the .05 level.

Table 12.6.   Psychiatric symptomatology by diagnosis
—results from case notes

Symptoms	Systemic Lupus Erythematosus		Rheumatoid Arthritis	
	N	(%)	N	(%)
Severe	5	(7%)	0	(0%)
Moderate	15	(22%)	2	(6%)
None	48	(71%)	34	(94%)
Total	68	(100%)	36	(100%)

A possible explanation of the difference just found, when the results from the interview are taken into account, is that a self-perpetuating myth may be in operation. As more and more reports are published indicating a high frequency of psychological disturbance in SLE, more and more clinicians will be influenced to observe and record its presence. In itself, this is not a bad thing. But, if care in recording such disturbances in SLE is at the expense of equal care in other disorders, then the scientific value of these observations becomes highly questionable. Differences between the two samples were in fact found, but they emerged from the structured interview data rather than from the case notes.

It therefore seems clear that an investigator should rely on a person ignorant of the status of the patient for the collection of background and other concomitant information. The desirability of a structured interview or questionnaire—where the questions to be asked and the probes to be made are set forth, and where the responses are precoded—may not be so clear.

Setting down the questions to be asked ensures that each interviewer covers the same ground, and that each subject is asked the same questions. In this way differences in interviewing style and biases due to different kinds of patients being asked different questions are reduced. Having the responses precoded serves not only to have the data in a form suitable for keypunching and machine processing, but also to reduce the errors in having a person make sense of a verbal, sometimes anecdotal report.

Such structured interviews have met with great success in psychiatry (Spitzer, Fleiss, Endicott, and Cohen, 1967; Wing et al., 1967; and Spitzer, Endicott, Fleiss, and Cohen, 1970) and in the study of respiratory diseases (Medical Research Council, 1966). The need for such procedures in psychiatry and bronchopulmonary medicine is due mainly to the variability among clinicians in the way they elicit information from patients, and, having elicited the data, in the way they interpret their findings.

Similar factors are present in almost every branch of medicine, and there is no compelling reason why the idea of the structured interview cannot be extended. Its applicability is clear in the interview for history, but rather subtle in the assessment of X-ray negatives, electrocardiogram tracings, and the like. An important reason for diagnostic disagreement in heart disease, for example, is that different cardiologists interpret EKGs differently; even the same cardiologist may, on two occasions, interpret the same EKG differently. Surely many of these differences would be reduced if cardiologists were instructed to note, on a precoded form, just what the abnormalities were that they thought they detected in each wave of the EKG. The same idea can certainly be applied to the recording of lesions thought to be detected from an X-ray negative.

Having information recorded on such forms, in addition to increasing uniformity, serves yet another purpose. The data, provided they are suitably recorded, can be quantified and can thus serve to provide a more objective gradation of disease severity than that based on clinical judgment. The major difficulty in the use of such a form, assuming that someone has the fortitude and patience to develop, validate, and if necessary revise one, is that clinicians would be unfamiliar with it and might resent having to use it. Considering, however, that the use of such techniques brings medical and epidemiological research closer to the ideal of all scientific endeavor—that all criteria be publicly specified and thus that every study be reproducible— not using them is virtually impossible to justify.

Problem 12.1. We considered, in Section 12.1, a means of correcting for the bias in an observed proportion. We now consider the comparison of two proportions, both of which are subject to error.

(a) The value of p_L, the rate of heavy smoking among lung cancer patients based on verbal report, was .44. Suppose that 200 women with cancer of the breast are interviewed, with 60 of them indicating that they are heavy smokers. What is the value of p_B, the rate of heavy smoking among breast cancer patients? What is the value of the odds ratio measuring the degree of association between type of cancer and heavy smoking?

(b) Based on the determination of smoking status by both response to interview and chemical test for 50 lung cancer patients, the rate of heavy smoking was adjusted to $P_L = .51$. Suppose that 50 breast cancer patients are likewise given a chemical test in addition to an interview, with the following results:

Chemical Test

Report	Heavy	Not Heavy	Total
Heavy	18	0	18
Not heavy	2	30	32

What are the values of the two correction factors, n_{00}/n_0 and n_{10}/n_1, which are to be applied to p_B? What value of P_B results from the adjustment

$$P_B = \frac{n_{00}}{n_0} p_B + \frac{n_{10}}{n_1} (1 - p_B)?$$

What is the value of the odds ratio associated with the adjusted rates? Does the association between heavy smoking and type of cancer now appear weaker or stronger than in (a)?

(c) Suppose that the cross-classification for the subsample of 50 breast cancer patients yielded the following results:

Chemical Test

Report	Heavy	Not Heavy	Total
Heavy	16	2	18
Not heavy	7	25	32

What are the values of the correction factors n_{00}/n_0 and n_{10}/n_1? What is the resulting value of P_B? What is the resulting value of the odds ratio? Does the association appear weaker or stronger than in (a)?

Problem 12.2. Suppose that the value of the correction factor n_{00}/n_0 was the same for the two kinds of patients, and that the value of the correction factor n_{10}/n_1 was likewise the same. What is a simple expression for $P_L - P_B$ as a function of $p_L - p_B$ and of the difference $n_{00}/n_0 - n_{10}/n_1$?

Problem 12.3. (a) Suppose that two raters rate each of 100 subjects, with the following results:

A

B	$+$	$-$	Total
$+$	30	10	40
$-$	30	30	60
Total	60	40	100

What is the value of kappa (12.13)? What is the value of its squared standard error (12.15)? How does the value of $\kappa^2/\text{Var}(\kappa)$ compare with the value of χ^2 for these data? (Do not use the continuity correction in calculating χ^2.)

(b) Suppose that two other raters rate each of another 100 subjects on the same characteristic, with the following results:

| | C | | |
D	+	−	Total
+	22	8	30
−	33	37	70
Total	55	45	100

What is the value of kappa? of its standard error? How does the value of $\kappa^2/\text{Var}(\kappa)$ compare with the value of χ^2 for these data? (Do not use the continuity correction.)

(c) What would you conclude about the difference in degree of *agreement* between the two pairs of raters? about the difference in degree of *association* between their ratings?

REFERENCES

Armitage, P., Blendis, L. M. and Smyllie, H. C. (1966). The measurement of observer disagreement in the recording of signs. *J. roy. statist. Soc., Ser. A*, **129**, 98–109.

Bartko, J. J. (1966). The intraclass correlation coefficient as a measure of reliability. *Psychol. Rep.*, **19**, 3–11.

Bryson, M. R. (1965). Errors of classification in a binomial population. *J. Amer. statist. Assoc.*, **60**, 217–224.

Burdock, E. I., Fleiss, J. L. and Hardesty, A. S. (1963). A new view of inter-observer agreement. *Personnel Psychol.*, **16**, 373–384.

Cochran, W. G. (1968). Errors of measurement in statistics. *Technometrics*, **10**, 637–666.

Cohen, J. (1960). A coefficient of agreement for nominal scales. *Educ. psychol. Meas.*, **20**, 37–46.

Cohen, J. (1968). Weighted kappa: Nominal scale agreement with provision for scaled disagreement or partial credit. *Psychol. Bull.*, **70**, 213–220.

Densen, P. M., Davidow, B., Bass, H. E. and Jones, E. W. (1967). A chemical test for smoking exposure. *Arch. environ. Health*, **14**, 865–874.

Ebel, R. L. (1951). Estimation of the reliability of ratings. *Psychometrika*, **16**, 407–424.

Everitt, B. S. (1968). Moments of the statistics kappa and weighted kappa. *Brit. J. math. statist. Psychol.*, **21**, 97–103.

Fleiss, J. L. (1963). Determination of the reliability of ratings by means of incomplete blocks designs. *Amer. Psychol.*, **18**, 420.

Fleiss, J. L. (1966). Assessing the accuracy of multivariate observations. *J. Amer. statist. Assoc.*, **61**, 403–412.

Fleiss, J. L. (1970). Estimating the reliability of interview data. *Psychometrika*, **35**, 143–162.

Fleiss, J. L. (1971). Measuring nominal scale agreement among many raters. *Psychol. Bull.*, **76**, 378–382.

Fleiss, J. L., Cohen, J. and Everitt, B. S. (1969). Large sample standard errors of kappa and weighted kappa. *Psychol. Bull.*, **72**, 323–327.

Fleiss, J. L., Spitzer, R. L. and Burdock, E. I. (1965). Estimating accuracy of judgment using recorded interviews. *Arch. gener. Psychiatr.*, **12**, 562–567.

Fleiss, J. L., Spitzer, R. L., Endicott, J. and Cohen, J. (1972). Quantification of agreement in multiple psychiatric diagnosis. *Arch. gener. Psychiatr.*, **26**, 168–171.

Ganz, V. H., Gurland, B. J., Deming, W. E. and Fisher, B. (1972). The study of the psychiatric symptoms of systemic lupus erythematosus: A biometric study. *Psychosom. Med.*, **34**, 199–206.

Guralnick, L. (1966). Some problems in the use of multiple causes of death. *J. chronic Dis.*, **19**, 979–990.

Haggard, E. A. (1958). *Intraclass correlation and the analysis of variance*. New York: Dryden Press.

Harper, D. (1964). Misclassification in epidemiological surveys. *Amer. J. public Health*, **54**, 1882–1886.

Krippendorff, K. (1970). Bivariate agreement coefficients for reliability of data. Pp. 139–150 in E. F. Borgatta (Ed.). *Sociological methodology 1970*. San Francisco: Jossey-Bass.

Levin, M. L. (1954). Etiology of lung cancer: Present status. *N. Y. State J. Med.*, **54**, 769–777.

Light, R. J. (1971). Measures of agreement for qualitative data: Some generalizations and alternatives. *Psychol. Bull.*, **76**, 365–377.

Maxwell, A. E. and Pilliner, A. E. G. (1968). Deriving coefficients of reliability and agreement for ratings. *Brit. J. math. statist. Psychol.*, **21**, 105–116.

Medical Research Council (1966). Questionnaire on respiratory diseases. Dawlish, Devon, England: W. J. Holman, Ltd.

Press, S. J. (1968). Estimating from misclassified data. *J. Amer. statist. Assoc.*, **63**, 123–133.

Rogot, E. and Goldberg, I. D. (1966). A proposed index for measuring agreement in test-retest studies. *J. chronic Dis.*, **19**, 991–1006.

Rubin, T., Rosenbaum, J. and Cobb, S. (1956). The use of interview data for the detection of association in field studies. *J. chronic Dis.*, **4**, 253–266.

Spitzer, R. L., Cohen, J., Fleiss, J. L. and Endicott, J. (1967). Quantification of agreement in psychiatric diagnosis. *Arch. gener. Psychiatr.*, **17**, 83–87.

Spitzer, R. L., Endicott, J., Fleiss, J. L. and Cohen, J. (1970). Psychiatric Status Schedule: A technique for evaluating psychopathology and impairment in role functioning. *Arch. gener. Psychiatr.*, **23**, 41–55.

Spitzer, R. L., Fleiss, J. L., Endicott, J. and Cohen, J. (1967). Mental Status Schedule: Properties of factor analytically derived scales. *Arch. gener. Psychiatr.*, **16**, 479–493.

Tenenbein, A. (1970). A double sampling scheme for estimating from binomial data with misclassifications. *J. Amer. statist. Assoc.*, **65**, 1350–1361.

Tenenbein, A. (1971). A double sampling scheme for estimating from binomial data with misclassifications: Sample size determination. *Biometrics*, **27**, 935–944.

Wing, J. K., Birley, J. L. T., Cooper, J. E., Graham, P. and Isaacs, A. D. (1967). Reliability of a procedure for measuring and classifying "present psychiatric state." *Brit. J. Psychiatr.*, **113**, 499–515.

CHAPTER 13

The Standardization of Rates

One of the most frequently occurring problems in epidemiology and vital statistics is the comparison of the rate for some event or characteristic across different populations or for the same population over time. If the populations were similar with respect to factors associated with the event under study—factors such as age, sex, race, or marital status—there would be no problem in comparing the overall rates (synonyms are total or crude rates) as they stand.

If the populations are not similarly constituted, however, the direct comparison of the overall rates may be misleading. Algebraically, the problem with such a comparison is as follows. Let $p_1, \ldots, p_I$ denote the proportions of all members of one of the populations being compared who fall into the various strata (age intervals, socioeconomic groups, etc.), there being I strata in all. Thus, $\sum p_i = 1$. If c_i denotes the rate specific to the ith stratum of this population, then the overall or crude rate for it is

$$c = \sum_{i=1}^{I} c_i p_i. \tag{13.1}$$

If the distribution for the second population across the I strata is represented by the proportions $P_1, \ldots, P_I$, so that $\sum P_i = 1$, and if C_i denotes the rate specific to the ith stratum in the second population, then the crude rate for this population is

$$C = \sum_{i=1}^{I} C_i P_i. \tag{13.2}$$

The difference between the two crude rates is

$$d = c - C, \tag{13.3}$$

and it is easy to check that

$$d = \sum_{i=1}^{I} \frac{p_i + P_i}{2} (c_i - C_i) + \sum_{i=1}^{I} \frac{c_i + C_i}{2} (p_i - P_i) \tag{13.4}$$

(see Kitagawa, 1955, and Hemphill and Ament, 1970).

155

It is thus seen that the difference between two crude rates has two components. One of them,

$$d_1 = \sum_{i=1}^{I} \frac{p_i + P_i}{2} (c_i - C_i), \tag{13.5}$$

is a true summarization of the differences between the two schedules of specific rates, differences that are usually of major interest. However, the second component,

$$d_2 = \sum_{i=1}^{I} \frac{c_i + C_i}{2} (p_i - P_i), \tag{13.6}$$

is a summarization of the differences between the two sets of population distributions, differences that are of little if any interest.

A number of conclusions can be drawn from the representation (13.4):

1. If the two population distributions are equal, that is, if $p_1 = P_1, \ldots,$ $p_I = P_I$, then $d_2 = 0$ and the difference between the crude rates indeed summarizes the differences between the schedules of specific rates.

2. If the two schedules of specific rates are equal, that is, if $c_1 = C_1, \ldots, c_I = C_I$, then $d_1 = 0$ and the difference between the crude rates measures only the difference in population distributions across the strata, a difference usually of no importance.

3. For any given value of d_1, that is, for a given summarization of the differences between two schedules of specific rates, the apparent difference between the two populations measured by d may be unaltered, increased, decreased, or even changed in sign depending on the differences between the two population distributions. The effect of d_2 is an additive one if the first population has a larger proportion of its members in the strata where the rates are high than does the second—that is, if $p_i > P_i$ in those strata where $(c_i + C_i)/2$ is large—and the effect of d_2 is a subtractive one if the converse holds. It is these kinds of effects that explain why, when each of one population's specific rates is greater than another's, the first population's crude rate may nevertheless be lower than the second's.

Section 13.1 presents some reasons for standardization, and some warnings against its uncritical use. Section 13.2 describes the indirect method of standardization, and Section 13.3 illustrates how it can give misleading results. Section 13.4 describes the direct method of standardization, Section 13.5 presents some other methods of standardization, and Section 13.6 discusses techniques for standardizing on two correlated dimensions.

13.1. REASONS FOR AND WARNINGS AGAINST STANDARDIZATION

It is to prevent anomalies of the sort described in (2) and (3) above that resort is made to standardization (synonymously, adjustment) in the comparison of two or more schedules of specific rates. Standardization should never, however, substitute for a comparison of the specific rates themselves. It is these that characterize the experience (morbidity, mortality, or whatever the rate refers to) of the populations being studied.

Woolsey has pointed out that "The specific rates are essential (because) it is only through the analysis of specific rates that an accurate and detailed study can be made of the variation (of the phenomenon under study) among population classes [1959, p. 60]." Elveback (1966), too, stresses the importance of studying the specific rates, and strongly criticizes the calculation of adjusted rates.

One criticism of the adjustment of rates is that if the specific rates vary in different ways across the various strata, then no single method of standardization will indicate that these differences exist. Standardization will, on the contrary, tend to mask these differences. As one example (see Kitagawa, 1966), there is the contrast between the age-specific death rates for white males resident in metropolitan counties of the United States in 1960 and those for white males resident in nonmetropolitan counties. Up to age 40, the rates in metropolitan counties are lower than in nonmetropolitan counties; after age 40, the reverse is true. No single summary comparison will reveal this information. On the contrary, at least two summary comparisons are needed.

Another example is provided by data reported by El-Badry (1969). He points out that in Ceylon, India, and Pakistan, mortality among males occurs at a lower rate than among females in many age categories. Single summary indexes for males and females might mask this phenomenon, and thus fail to reveal data suggestive of further research. Doll and Cook (1967), in addition, cite the inadequacy of a single index for summarizing age- and sex-specific incidence rates.

Bearing in mind that there is no substitute for examining the specific rates themselves, we may consider some of the reasons for standardization.

1. A single summary index for a population is more easily compared with other summary indices than are entire schedules of specific rates.

2. If some strata are comprised of small numbers of people, the associated specific rates may be too imprecise and unreliable for use in detailed comparisons.

3. For small populations, or for some groups of especial interest, specific rates may not exist. This may be the case for selected occupational groups and for populations from geographic areas especially demarcated for a single study. In such cases, only the total number of events (e.g., deaths) may be available, and not their subdivision by strata.

Other reasons for standardization are given by Woolsey (1959), Kalton (1968), and Cochran (1968), who in addition studied the effects of varying the number of strata, I.

13.2. INDIRECT STANDARDIZATION

The second and third reasons just given for standardization, the unreliability and possibly even the unavailability of some specific rates, lead to perhaps the most frequently adopted method of standardization, the so-called indirect method. The ingredients necessary for its implementation are:

1. The crude rate for the population being studied, say c.
2. The distribution across the various strata for that population, say $p_1, \ldots, p_I$.
3. The schedule of specific rates for a selected standard population, say $c_{S1}, \ldots, c_{SI}$.
4. The crude rate for the standard population, say c_S.

The first calculation in indirect standardization is of the overall rate that would obtain if the schedule of specific rates for the standard population were applied to the given population. It is

$$c' = \sum_{i=1}^{I} c_{Si} p_i. \tag{13.7}$$

The indirect adjusted rate is then

$$c_{\text{indirect}} = c_S \frac{c}{c'} ; \tag{13.8}$$

that is, the crude rate for the standard population, c_S, multiplied by the ratio of the actual crude rate for the given population, c, to the crude rate, c', which would exist if the given population were subject to the standard population's schedule of rates.

As an example, consider the following data from Stark and Mantel (1966). In the state of Michigan, from 1950 to 1964, 731,177 infants were the firstborn to their mothers; of these, 412 were mongoloid, giving a crude rate of $c = 56.3$ mongoloids per 100,000 firstborn live births. In the same 15-year interval, 442,811 infants were the fifth-born or more to their mothers; of these, 740 were mongoloid, giving a crude rate of $C = 167.1$ mongoloids per 100,000 fifth-born or more live births.

The comparison as it stands is not a fair one, because maternal age is known to be associated with both birth order and mongolism. Some method has therefore to be applied to adjust for possible differences between the firstborn and later-born in maternal age distributions. Table 13.1 illustrates indirect adjustment.

Table 13.1. An example of indirect standardization

| Maternal Age | Specific Rates for all Michigan per 100,000—c_{Si} | Birth Order | | | |
| | | First | | Fifth or More | |
		p_i	$c_{Si}p_i$	P_i	$c_{Si}P_i$
Under 20	42.5	.315	13.4	.001	0.0
20–24	42.5	.451	19.2	.069	2.9
25–29	52.3	.157	8.2	.279	14.6
30–34	87.7	.054	4.7	.339	29.7
35–39	264.0	.019	5.0	.235	62.0
40 and over	864.4	.004	3.5	.078	67.4
Sum			54.0 ($= c'$)		176.6 ($= C'$)

The selected standard population was all the live births in Michigan during the years 1950–1964. The crude rate of mongolism for the state as a whole was $c_S = 89.5$ per 100,000 live births, and the rates specific for maternal age are given in the column headed c_{Si} in the table. For the infants born first and born fifth or more, the maternal age distributions are given in the columns headed p_i and P_i. The results of applying formula (13.7) are shown in the bottom row of the table.

To review, we are given the crude rates

$$c = 56.3 \text{ mongoloids per 100,000 firstborns} \qquad (13.9)$$

and

$$C = 167.1 \text{ mongoloids per 100,000 fifth- or later-borns.} \qquad (13.10)$$

By applying the rates of mongolism specific to maternal age for the state of Michigan as a whole, we would have expected

$$c' = 54.0 \text{ mongoloids per 100,000 firstborns,}$$

and

$$C' = 176.6 \text{ mongoloids per 100,000 fifth- or later-borns.}$$

Given the crude rate for the entire state,

$$c_S = 89.5 \text{ mongoloids per 100,000 live births,}$$

we find, by (13.8), the indirect adjusted rates

$$c_{\text{indirect}} = 89.5 \times \frac{56.3}{54.0}$$

$$= 93.3 \text{ mongoloids per } 100{,}000 \text{ firstborns} \qquad (13.11)$$

and

$$C_{\text{indirect}} = 89.5 \times \frac{167.1}{176.6}$$

$$= 84.7 \text{ mongoloids per } 100{,}000 \text{ fifth- or later-borns.} \quad (13.12)$$

By just comparing the crude rates given in (13.9) and (13.10), we would conclude that there was a threefold increase in the risk of mongolism from firstborns to infants born fifth or more. By comparing the adjusted rates given by (13.11) and (13.12), on the other hand, we would conclude that there was no effective difference in the risk of mongolism.

It seems that the apparent greater risk of mongolism for later births, suggested by a comparison of the crude rates (13.9) and (13.10), is a reflection of differences in maternal age distribution. Proportionately more mothers of later-born infants are in the older age categories, where the specific rates are higher, than are mothers of firstborn infants. After adjustment for differences in the maternal age distributions, it appears that, if anything, the rate of mongolism in later-born infants is somewhat less than the rate in firstborn infants.

13.3. A FEATURE OF INDIRECT STANDARDIZATION

Consider the data of Table 13.2, giving hypothetical sex-specific mortality rates (per 1000) in each of two groups. The two sets of sex-specific rates are

Table 13.2. Sex-specific mortality rates in two groups

Sex	Group 1		Group 2	
	p_i	Rate/1000	P_i	Rate/1000
Male	.60	2.0	.80	2.0
Female	.40	1.0	.20	1.0

equal, but the unequal sex distributions in the two groups yield unequal overall rates. For group 1, the crude rate is

$$c = 2.0 \times .60 + 1.0 \times .40 = 1.6 \text{ deaths}/1000, \qquad (13.13)$$

and for group 2 the crude rate is

$$C = 2.0 \times .80 + 1.0 \times .20 = 1.8 \text{ deaths}/1000. \tag{13.14}$$

Suppose, now, that the two sets of sex-specific rates could have been obtained only with the greatest difficulty, so that the only data actually available were the two sex distributions and the two crude rates. Indirect adjustment would therefore have to be resorted to. For the population chosen as the standard, suppose that the crude rate is

$$c_S = 1.5 \text{ deaths}/1000, \tag{13.15}$$

and that the sex-specific rates are

$$c_{S1} = 2.2 \text{ deaths}/1000 \text{ males} \tag{13.16}$$

and

$$c_{S2} = 0.9 \text{ deaths}/1000 \text{ females.} \tag{13.17}$$

The expected crude rate in group 1 is

$$c' = 2.2 \times .60 + 0.9 \times .40 = 1.68 \text{ deaths}/1000, \tag{13.18}$$

yielding an indirect adjusted rate of

$$c_{\text{indirect}} = 1.5 \times \frac{1.6}{1.68} = 1.43 \text{ deaths}/1000. \tag{13.19}$$

The expected crude rate in group 2 is

$$C' = 2.2 \times .80 + 0.9 \times .20 = 1.94 \text{ deaths}/1000, \tag{13.20}$$

so that, for group 2,

$$C_{\text{indirect}} = 1.5 \times \frac{1.8}{1.94} = 1.39 \text{ deaths}/1000. \tag{13.21}$$

The two adjusted rates given by (13.19) and (13.21) are more nearly equal than the two unadjusted rates given by (13.13) and (13.14), reflecting more accurately the equality of sex-specific rates indicated in Table 13.2. It is a bit disquieting, however, that equality of the two schedules of specific rates has not been reflected by precise equality of the two indirect adjusted rates. This is a feature of indirect adjustment that does not characterize direct adjustment, the method to be considered next. Neither in this example nor in most other instances is the distortion great, however. Furthermore, such distortion will not occur if the standard population is the composite of the two populations studied.

It is clear that indirect standardization does not completely adjust for differences in population composition. Thus, when attempting to explain variability across groups of indirect adjusted rates, one should bear in mind that, whereas variation of schedules of specific rates accounts for most of it, variation in population composition may still account for some of it. Additional criticisms of indirect adjustment are given by Yule (1934) and Kilpatrick (1963).

13.4. DIRECT STANDARDIZATION

The method of standardization used second most frequently is the so-called direct method. Direct standardization may be applied only when the schedule of specific rates for a given population is available. The data necessary for its implementation are:

1. The schedule of specific rates for the population being studied, say $c_1, \ldots, c_I$.
2. The distribution across the various strata for a selected standard population, say $p_{S1}, \ldots, p_{SI}$.

The direct adjusted rate is then simply

$$c_{\text{direct}} = \sum_{i=1}^{I} c_i p_{Si}. \tag{13.22}$$

The term direct refers to working directly with the specific rates of the population being studied, in distinction to what was done in the method previously presented.

As an example, let us consider the same event, mongolism, studied in Section 13.2. Table 13.3 gives the maternal age distribution for all infants born in the state of Michigan from 1950 to 1964 (the standard distribution), and the rates of mongolism specific to maternal age for firstborns and for infants born fifth or more (data from Stark and Mantel, 1966).

Table 13.3.   An example of direct standardization (rates per 100,000)

| | | Birth Order | | | |
| | Distribution for all of | First | | Fifth or More | |
Maternal Age	Michigan-p_{Si}	c_i	$c_i p_{Si}$	C_i	$C_i p_{Si}$
Under 20	.113	46.5	5.3	0	0.0
20–24	.330	42.8	14.1	26.1	8.6
25–29	.278	52.2	14.5	51.0	14.2
30–34	.173	101.3	17.5	74.7	12.9
35–39	.084	274.5	23.1	251.7	21.1
40 and over	.022	819.1	18.0	857.8	18.9
Sum			$\overline{92.5}$ $(= c_{\text{direct}})$		$\overline{75.7}$ $(= C_{\text{direct}})$

The conclusion drawn from a comparison of these direct adjusted rates is the same as that drawn before from the indirect adjusted rates: the rates of mongolism are about the same for infants born first and for infants born fifth or more. Such concordance between the conclusions drawn from comparisons of indirect and direct adjusted rates is usually, although not invariably, the case.

Given the consistent contrast between the specific rates shown in Table 13.3—in five of the six maternal age categories, the rate of mongolism among firstborns was slightly greater than the rate among infants born fifth or more—there was clearly no compelling reason for any standardization at all. The specific rates spoke for themselves. The only legitimate reason for standardization in such a case is the one cited first at the conclusion of Section 13.1, namely the greater simplicity of working with a single summary index than of working with an entire schedule of specific rates.

There is one decided advantage to direct over indirect standardization. If, stratum by stratum, the specific rate in one group is equal to the specific rate in a second group, then, no matter which population is chosen as standard, the direct adjusted rates will be equal. Consider, for example, the specific rates presented in Table 13.2. The direct adjusted rate for group 1 is

$$c_{\text{direct}} = 2.0 \times p_{S1} + 1.0 \times p_{S2}, \qquad (13.23)$$

and that for group 2 is obviously the same.

Direct standardization has a more general property. Consistent inequalities among specific rates, stratum by stratum, yield direct adjusted rates bearing the same inequalities. Thus if each specific rate in group 1 is greater than the corresponding rate in group 2, the direct adjusted rate for group 1 will be greater than that for group 2, no matter what the composition of the standard population.

These features of direct adjusted rates are actually quite trivial, for the circumstances leading to them are fully described by the specific rates themselves. Here the adjusted rates serve merely as convenient summarizations.

An important point to bear in mind is that an adjusted rate, no matter which method of adjustment is used, has meaning only when compared with a similarly adjusted rate. Its magnitude means little in and of itself. For the rates of Table 13.2, for example, the direct adjusted rate varies from 1.25 through 1.50 to 1.75 as the standard sex distribution varies from (.25, .75) through (.50, .50) to (.75, .25). No matter which standard is used, the direct adjusted rates for the two groups will be identical. The magnitude of the rate, however, is seen to depend strongly on the composition of the standard population. Spiegelman and Marks (1966) have shown that, in the direct standardization of mortality rates, the choice of a standard population generally has little effect on the differences between adjusted rates, and tends to affect only their individual magnitudes.

When the specific rates in the groups being compared do not bear consistent relations across the strata, then any kind of overall standardization is questionable. Problem 13.1 is concerned with the comparison of two groups for which the specific rates of one are higher than those for the other in some strata but are lower in other strata. It is shown that, depending on the strata in which the standard population is concentrated, either of the two groups can end up with the larger adjusted rate. The standard population may even be chosen to give equal adjusted rates. In addition, the phenomenon of a crossover in specific rates is lost by the usual methods of standardization.

A compromise calls for the calculation of a number of adjusted rates, one for each contiguous set of strata in which the specific rates bear consistent relations over the groups being compared. This device is illustrated in Problem 13.1, where two such sets of strata may be constituted. The division of the strata into such sets is not always easy, and sometimes may have to be forced. Nevertheless, working with a number of adjusted rates is preferable to working with one overall adjusted rate which may be more of a distortion than a summarization.

13.5. SOME OTHER SUMMARY INDICES

Woolsey (1959) and Kitagawa (1964) have reviewed a number of approaches to the standardization of rates. Formulas for determining standard errors have been given by Chiang (1961) and Keyfitz (1966).

Variations are sometimes encountered of the two kinds of adjusted rates considered so far. With respect to indirect standardization, the quantity

$$\text{SMF} = \frac{c}{c'} = \frac{c_{\text{indirect}}}{c_S} \tag{13.24}$$

may be used, where c_S is the rate for the standard population. This quantity, called the *standard mortality figure* when the event studied is mortality, is merely the ratio of the actual to the expected crude rate. The corresponding ratio for direct adjustment,

$$\text{CMF} = \frac{c_{\text{direct}}}{c_S}, \tag{13.25}$$

is dubbed the *comparative mortality figure* when applied to mortality.

Some other methods of adjustment exist, but are less frequently used than the ones so far studied. One is a simple average of the crude and the direct adjusted rates,

$$CMR = \tfrac{1}{2}(c + c_{\text{direct}}) = \sum_{i=1}^{I} \tfrac{1}{2}(p_{Si} + p_i)c_i, \tag{13.26}$$

and is referred to as the *comparative mortality rate* when mortality is studied. Its infrequent use is testimony to its virtual uninterpretability.

Two indices are available for use with age-specific rates which in effect give equal weight to each year of age. Let n_i denote the number of years in the ith age interval, so that if, for example, the first age interval is 0–4 years, then $n_1 = 5$. The first such index (see Yule, 1934) is

$$EADR = \frac{\sum_{i=1}^{I} n_i c_i}{\sum_{i=1}^{I} n_i}, \tag{13.27}$$

named the *equivalent average death rate* when applied to mortality. This index may be viewed as a direct adjusted rate where each year of age is assumed to have the same number of people.

The second such index (see Yerushalmy, 1951, and Elveback, 1966) is

$$MI = \frac{\sum_{i=1}^{I} n_i \dfrac{c_i}{c_{Si}}}{\sum_{i=1}^{I} n_i}, \tag{13.28}$$

named the *mortality index* when applied to mortality. The mortality index is a simple average of the ratios of specific rates, with weights given by the numbers of years in the various age intervals. The usefulness of the latter two indices is limited because of the questionable validity of assigning equal importance to each year of age.

An index similar to (13.28) is

$$RMI = \sum_{i=1}^{I} p_i \frac{c_i}{c_{Si}}, \tag{13.29}$$

named the *relative mortality index* when applied to mortality. The relative mortality index is also an average of the ratios of the actual specific rates to the standard population's specific rates, but is weighted by the given population's actual age distribution.

An equivalent expression is

$$RMI = \frac{\sum_{i=1}^{I} \dfrac{e_i}{c_{Si}}}{N}, \tag{13.30}$$

where e_i is the observed *number* of deaths (in general, of events) in the ith stratum and N is the total number of people in the given population. It is thus seen that one only needs, for the given population, its total size and the distribution over its strata of its total number of events in order to calculate the relative mortality index. An implication is that the relative mortality index may be calculated in years between censuses when, say, the age distribution of the population is not available but when the age distribution of deaths may be determined from registration data.

13.6. ADJUSTMENT FOR TWO FACTORS

Table 13.4 presents data on the incidence of mongolism specific both to birth order and to maternal age. The two methods to be described, the first based on direct and the second on indirect standardization, are useful when two factors are both associated with some disorder and with each other, and when one wishes to identify and measure their separate effects.

Table 13.4. *Distribution of discovered mongoloids and of total live births by maternal age and birth order, Michigan, 1950–1964 (Values in cells are number of mongoloids found per number of live births)**

| | Birth Order | | | | | |
Maternal Age	1	2	3	4	5+	Total
Under 20	$\frac{107}{230,061}$	$\frac{25}{72,202}$	$\frac{3}{15,050}$	$\frac{1}{2293}$	$\frac{0}{327}$	$\frac{136}{319,933}$
20–24	$\frac{141}{329,449}$	$\frac{150}{326,701}$	$\frac{71}{175,702}$	$\frac{26}{68,800}$	$\frac{8}{30,666}$	$\frac{396}{931,318}$
25–29	$\frac{60}{114,920}$	$\frac{110}{208,667}$	$\frac{114}{207,081}$	$\frac{64}{132,424}$	$\frac{63}{123,419}$	$\frac{411}{786,511}$
30–34	$\frac{40}{39,487}$	$\frac{84}{83,228}$	$\frac{103}{117,300}$	$\frac{89}{98,301}$	$\frac{112}{149,919}$	$\frac{428}{488,235}$
35–39	$\frac{39}{14,208}$	$\frac{82}{28,466}$	$\frac{108}{45,026}$	$\frac{137}{46,075}$	$\frac{262}{104,088}$	$\frac{628}{237,863}$
40 and over	$\frac{25}{3052}$	$\frac{39}{5375}$	$\frac{75}{8660}$	$\frac{96}{9834}$	$\frac{295}{34,392}$	$\frac{530}{61,313}$
Total	$\frac{412}{731,177}$	$\frac{490}{724,639}$	$\frac{474}{568,819}$	$\frac{413}{357,727}$	$\frac{740}{442,811}$	$\frac{2529}{2,825,173}$

* Data from Stark and Mantel (1966).

Simultaneous Direct Adjustment. Table 13.5 presents the various specific rates, the crude rates, and the direct adjusted rates.

Table 13.5. *Incidence rates of discovered mongolism (number of mongoloids found per 100,000 live births) by maternal age and birth order*

Maternal Age	Birth Order					Crude Rate	Adjusted Rate†
	1	2	3	4	5+		
Under 20	46.5	34.6	19.9	43.6	0	42.5	30.4
20–24	42.8	45.9	40.4	37.8	26.1	42.5	39.9
25–29	52.2	52.7	55.1	48.3	51.0	52.3	52.2
30–34	101.3	100.9	87.8	90.5	74.7	87.7	92.9
35–39	274.5	288.1	239.9	297.3	251.7	264.0	270.3
40 and over	819.1	725.6	866.1	976.2	857.8	864.4	830.4
Crude rate	56.3	67.6	83.3	115.5	167.1	89.5	
Adjusted rate*	92.3	91.2	85.1	92.7	75.5		88.0‡

* The last row contains rates specific for birth order and directly adjusted for maternal age, with the standard maternal age distribution being that of the total sample.

† The last column contains rates specific for maternal age and directly adjusted for birth order, with the standard birth-order distribution being that of the total sample.

‡ Whenever the two standard distributions are those of the total sample, the overall rates based on the two series of directly adjusted rates will be equal to each other, but not necessarily to the overall crude rate.

In this case the specific rates speak for themselves. Within none of the maternal age categories is there any appreciable variation in the rates of mongolism specific to birth order. The increasing gradient with birth order of the crude rates is therefore likely to be only a reflection of the association between birth order and maternal age, and not of any direct relationship between birth order and the incidence of mongolism.

Maternal age, on the other hand, is seen to be strongly associated with the incidence of mongolism. Within each birth-order category, there is a clear increase in the incidence rate with increasing maternal age.

The direct adjusted rates shown in the last row and last column of Table 13.5 serve only to summarize the information provided by the 30 specific rates: little effect of birth order but a strong effect of maternal age on the incidence of mongolism. In fact, direct adjustment is appropriate only because of the consistency found both within maternal age categories (rather little variability among the birth-order-specific rates) and within birth-order categories (a clear gradient with increasing maternal age).

Simultaneous Indirect Adjustment. If all the specific rates are available, and all are based on samples of sufficient size to possess adequate precision, then simultaneous direct adjustment, the method just described, may be applied. If the specific rates are not available or are based on relatively small sample sizes, then a method due to Mantel and Stark (1968), based on indirect adjustment, may be applied.

It might have happened that the data required to calculate rates specific for maternal age and birth order simultaneously did not exist. In fact, the only data available might have been those in Table 13.6, together with the crude rates for birth order and for maternal age.

Table 13.6. *Distribution of total live births by maternal age and birth order, overall crude rates, and indirectly adjusted rates*

Maternal Age	Birth Order					Crude Rate	Adjusted Rate†
	1	2	3	4	5+		
Under 20	230,061	72,202	15,050	2,293	327	42.5	62.7
20–24	329,449	326,701	175,702	68,800	30,666	42.5	51.9
25–29	114,920	208,667	207,081	132,424	123,419	52.3	49.9
30–34	39,487	83,228	117,300	98,301	149,919	87.7	70.9
35–39	14,208	28,466	45,026	46,075	104,088	264.0	192.6
40 and over	3,052	5,375	8,660	9,834	34,392	864.4	582.9
Crude rate	56.3	67.6	83.3	115.5	167.1	89.5	79.2‡
Adjusted rate*	93.0	92.7	87.3	94.3	84.8	90.7‡	

* The last row contains rates specific for birth order and indirectly adjusted for maternal age, with the standard set of rates specific for maternal age being that of the total sample. Thus, for example,

$$c_{2(\text{indirect})} = 92.7 = 89.5 \times \frac{67.6 \times 7.24639}{42.5 \times .72202 + \cdots + 864.4 \times .05375}.$$

† The last column contains rates specific for maternal age and indirectly adjusted for birth order, with the standard set of rates specific for birth order being that of the total sample. Thus, for example,

$$c_{20-24(\text{indirect})} = 51.9 = 89.5 \times \frac{42.5 \times 9.31318}{56.3 \times 3.29449 + \cdots + 167.1 \times .30666}.$$

‡ The overall rates based on the two series of indirectly adjusted rates will almost never equal each other, nor will either of them equal the overall crude rate.

In contrast to the summarizing role played by the direct adjusted rates, the indirect adjusted rates, presented in the last row and last column of Table 13.6, must almost of necessity be calculated when the schedules of specific rates do not exist. They must be interpreted with a great deal of caution, however.

Two awkward features of the indirectly adjusted rates can be seen. For one thing, they do not yield equal overall rates. For another, the indirectly adjusted rates for some of the maternal age categories are totally out of the range of the specific rates (compare, e.g., the indirect adjusted rate of 62.7 for the first maternal age category with the specific rates ranging from 0 to 46.5, as seen in Table 13.5).

As a means of correcting these and possibly other anomalies, Mantel and Stark (1968) recommend the following procedure.

1. Beginning with the schedule of crude rates specific to maternal age, obtain the indirect adjusted rates specific for birth order. These have already been calculated, and appear in the final row of Table 13.6.

2. Using *these latter rates* as the standard schedule, calculate the indirect adjusted rates specific for maternal age. Multiply the obtained rates by the ratio of the expected total crude rate (90.7) to the actual total crude rate (89.5):

Maternal age	Under 20	20–24	25–29	30–34	35–39	40 and Over
Adjusted rate	41.6	42.0	52.4	89.0	270.3	892.9

For example, 41.6 is calculated as

$$41.6 = \frac{42.5}{92.7} \times 89.5 \times \frac{90.7}{89.5} = \frac{42.5 \times 90.7}{92.7},$$

where 42.5 is the crude rate in the first maternal age category, 90.7 is the overall expected crude rate, and

$$92.7 = \frac{93.0 \times 2.30061 + 92.7 \times .72202 + \cdots + 84.8 \times .00327}{3.19933}.$$

3. Continue in the same manner, each time using the previous set of indirect adjusted rates for one of the two variables to generate a new set for the other variable. Multiply each rate in the new set by the ratio of the expected total crude rate based on the previous set to the actual total crude rate.

4. Stop the process when successive sets of rates are unchanged. Here, about four cycles were necessary.

5. The sets of rates we end up with are

Maternal age	Under 20	20–24	25–29	30–34	35–39	40 and Over
Adjusted rate	41.0	41.8	52.6	89.7	273.3	904.5

for maternal age and

Birth order	1	2	3	4	5+
Adjusted rate	95.2	93.7	87.3	93.6	83.6

for birth order.

6. These two schedules of rates have the property that either, when used as the standard, implies the other. Furthermore, each of these rates happens for these data to lie within the range of the specific rates. This need not always occur with the Mantel-Stark adjustment method. Finally, their adjusted rates both yield the same expected total rate, 91.2. This final property would not have held had we not, at every step, multiplied the generated set of rates by the ratio of the total expected rate based on the preceding set of rates to the total observed rate.

The inferences from these adjusted rates are the same as were drawn previously: a strong effect of maternal age and little if any effect of birth order. Recall, however, that here we have made no use at all of the specific rates.

The Mantel-Stark procedure has the property that it will yield the same results no matter what set of rates one starts with, although the number of steps will vary with the starting set. An alternative approach to the problem of separating the effects of two correlated factors is to use one of the statistical models reviewed in Section 10.7. One of those models was actually applied by Berry (1970) to these data, with no change in inferences. Berry's analysis has the advantages of being generalizable to more than two factors and of providing tests of significance. The required arithmetic, however, is far more complex than for the Mantel-Stark method.

Problem 13.1. The following data are from Table 2 of Discher and Feinberg (1969).

Age-specific rates of abnormal lung functioning in males employed in manufacturing or service industries

Age Interval	Manufacturing		Services	
	Number	% Abnormal	Number	% Abnormal
20–29	403	2.2	256	4.8
30–39	688	3.2	525	3.2
40–49	683	2.2	599	2.8
50–59	539	6.9	453	6.6
60+	133	12.8	155	9.0

(a) Comparing the age-specific rates for the two kinds of industries, what conclusions would you be willing to draw?

(b) What do you think would be gained by calculating an age-adjusted rate for each kind of industry and then comparing the adjusted rates? What do you think might be lost?

(c) Consider the three following standard age distributions.

Standard

Age Interval	1	2	3
20–29	.25	.05	.07
30–39	.25	.05	.75
40–49	.30	.10	.06
50–59	.10	.40	.06
60+	.10	.40	.06

(1) The first standard distribution is concentrated below age 49. What are the values of the two direct adjusted rates using this standard? What is the direction and magnitude of difference between them?

(2) The second standard distribution is concentrated above age 50. What are the values of the two direct adjusted rates using this standard? What is the direction and magnitude of difference between them?

(3) The third standard distribution is concentrated in the age interval 30–39. What are the values of the two direct adjusted rates using this standard? What is the direction and magnitude of difference between them?

(d) Using the age distribution of the total sample as a standard, calculate two direct adjusted rates for both kinds of industries, one for the age interval 20–49 and the other for the interval 50 and over. Do the comparisons now seem fair to the data?

REFERENCES

Berry, G. (1970). Parametric analysis of disease incidences in multiway tables. *Biometrics*, **26**, 572–579.

Chiang, C. L. (1961). Standard error of the age-adjusted death rate. U.S. Department of Health, Education and Welfare: Vital Statistics—Special Reports, **47**, 271–285.

Cochran, W. G. (1968). The effectiveness of adjustment by subclassification in removing bias in observational studies. *Biometrics*, **24**, 295–313.

Discher, D. P. and Feinberg, H. C. (1969). Screening for chronic pulmonary disease: Survey of 10,000 industrial workers. *Amer. J. public Health*, **59**, 1857–1867.

Doll, R. and Cook, P. (1967). Summarizing indices for comparison of cancer incidence data. *Internat. J. Cancer*, **2**, 269–279.

El-Badry, M. A. (1969). Higher female than male mortality in some countries of south Asia: A digest. *J. Amer. statist. Assoc.*, **64**, 1234–1244.

Elveback, L. R. (1966). Discussion of "Indices of mortality and tests of their statistical significance." *Hum. Biol.*, **38**, 322–324.

Hemphill, F. M. and Ament, R. P. (1970). Quantitative assessment of subcategory contributions to observed change in relative frequencies. Paper read at annual meeting of American Public Health Association, Houston.

Kalton, G. (1968). Standardization: A technique to control for extraneous variables. *Appl. Statist.*, **17**, 118–136.

Keyfitz, N. (1966). Sampling variance of standardized mortality rates. *Hum. Biol.*, **38**, 309–317.

Kilpatrick, S. J. (1963). Mortality comparisons in socio-economic groups. *Appl. Statist.*, **12**, 65–86.

Kitagawa, E. M. (1955). Components of a difference between two rates. *J. Amer. statist. Assoc.*, **50**, 1168–1194.

Kitagawa, E. M. (1964). Standardized comparisons in population research. *Demography*, **1**, 296–315.

Kitagawa, E. M. (1966). Theoretical considerations in the selection of a mortality index, and some empirical comparisons. *Hum. Biol.*, **38**, 293–308.

Mantel, N. and Stark, C. R. (1968). Computation of indirect-adjusted rates in the presence of confounding. *Biometrics*, **24**, 997–1005.

Spiegelman, M. and Marks, H. H. (1966). Empirical testing of standards for the age adjustment of death rates by the direct method. *Hum. Biol.*, **38**, 280–292.

Stark, C. R. and Mantel, N. (1966). Effects of maternal age and birth order on the risk of mongolism and leukemia. *J. natl. Cancer Inst.*, **37**, 687–698.

Woolsey, T. D. (1959). Adjusted death rates and other indices of mortality. Chapter 4 in F. E. Linder and R. D. Grove. *Vital statistics rates in the United States, 1900–1940.* Washington, D.C.: U.S. Government Printing Office.

Yerushalmy, J. (1951). A mortality index for use in place of the age-adjusted death rate. *Amer. J. public Health*, **41**, 907–922.

Yule, G. U. (1934). On some points relating to vital statistics, more especially statistics of occupation mortality. *J. roy. statist. Soc.*, **97**, 1–72.

Appendix

TABLE A.1. CRITICAL VALUES OF THE CHI SQUARE DISTRIBUTION

DEGREES OF FREEDOM	ALPHA					
	.10	.05	.025	.01	.005	.001
1	2.71	3.84	5.02	6.63	7.88	10.83
2	4.61	5.99	7.38	9.21	10.60	13.82
3	6.25	7.81	9.35	11.34	12.84	16.27
4	7.78	9.49	11.14	13.28	14.86	18.47
5	9.24	11.07	12.83	15.09	16.75	20.52
6	10.64	12.59	14.45	16.81	18.55	22.46
7	12.02	14.07	16.01	18.48	20.28	24.32
8	13.36	15.51	17.53	20.09	21.96	26.12
9	14.68	16.92	19.02	21.67	23.59	27.88
10	15.99	18.31	20.48	23.21	25.19	29.59
11	17.28	19.68	21.92	24.72	26.76	31.26
12	18.55	21.03	23.34	26.22	28.30	32.91
13	19.81	22.36	24.74	27.69	29.82	34.53
14	21.06	23.68	26.12	29.14	31.32	36.12
15	22.31	25.00	27.49	30.58	32.80	37.70
16	23.54	26.30	28.85	32.00	34.27	39.25
17	24.77	27.59	30.19	33.41	35.72	40.79
18	25.99	28.87	31.53	34.81	37.16	42.31
19	27.20	30.14	32.85	36.19	38.58	43.82
20	28.41	31.41	34.17	37.57	40.00	45.32
25	34.38	37.65	40.65	44.31	46.93	52.62
30	40.26	43.77	46.98	50.89	53.67	59.70
40	51.80	55.76	59.34	63.69	66.77	73.40
60	74.40	79.08	83.30	88.38	91.95	99.61
100	118.50	124.34	129.56	135.81	140.17	149.45

ABRIDGED FROM TABLE 8 OF ''BIOMETRIKA TABLES FOR STATISTICIANS, VOL. I, 2ND EDITION,'' EDITED BY E.S. PEARSON AND H.O. HARTLEY. CAMBRIDGE UNIVERSITY PRESS, CAMBRIDGE, ENGLAND, 1958.

TABLE A.2. CRITICAL VALUES OF THE NORMAL DISTRIBUTION

P = AREA IN THE TAILS OF THE NORMAL CURVE BELOW −Z AND ABOVE +Z. IN A TEST OF SIGNIFICANCE, P IS THE SIGNIFICANCE LEVEL ASSOCIATED WITH THE OBTAINED VALUE OF Z.

THE TOTAL AREA UNDER THE NORMAL CURVE TO THE RIGHT OF Z IS 1 − P/2 IF Z IS NEGATIVE, AND IS P/2 IF Z IS POSITIVE.

SUPPOSE ONE MUST FIND THAT VALUE OF Z SUCH THAT THE TOTAL AREA UNDER THE NORMAL CURVE TO THE RIGHT OF Z IS 1 − B. IF 1 − B IS GREATER THAN 0.50, TAKE THE VALUE OF Z CORRESPONDING TO P = 2B, AND AFFIX A MINUS SIGN TO IT. IF 1 − B IS LESS THAN 0.50, TAKE THE VALUE OF Z CORRESPONDING TO P = 2(1 − B).

Z	P	Z	P	Z	P
0.0	1.0000	1.2	0.2301	2.4	0.0164
0.1	0.9203	1.282	0.20	2.5	0.0124
0.126	0.90	1.3	0.1936	2.576	0.01
0.2	0.8415	1.4	0.1615	2.6	0.0093
0.3	0.7642	1.440	0.15	2.7	0.0069
0.385	0.70	1.5	0.1336	2.8	0.0051
0.4	0.6892	1.6	0.1096	2.813	0.005
0.5	0.6171	1.645	0.10	2.9	0.0037
0.524	0.60	1.7	0.0891	3.0	0.0027
0.6	0.5485	1.8	0.0719	3.090	0.002
0.674	0.50	1.9	0.0574	3.1	0.0019
0.7	0.4839	1.960	0.05	3.2	0.0014
0.8	0.4237	2.0	0.0455	3.3	0.0010
0.842	0.40	2.1	0.0357	3.4	0.0007
0.9	0.3681	2.2	0.0278	3.5	0.0005
1.0	0.3173	2.242	0.025	3.6	0.0003
1.036	0.30	2.3	0.0214	3.7	0.0002
1.1	0.2713	2.326	0.02	3.8	0.0001

ADAPTED FROM TABLES 1 AND 4 OF ''BIOMETRIKA TABLES FOR STATISTICIANS, VOL. I, 2ND EDITION,'' EDITED BY E.S. PEARSON AND H.O. HARTLEY. CAMBRIDGE UNIVERSITY PRESS, CAMBRIDGE, ENGLAND, 1958.

TABLE A.3. SAMPLE SIZES PER GROUP FOR TWO-TAILED TEST ON PROPORTIONS
P1 = 0.05

P2	ALPHA	0.99	0.95	0.90	0.85	POWER 0.80	0.75	0.70	0.65	0.50
0.10	0.01	1407.	1064.	902.	800.	725.	663.	610.	563.	445.
0.10	0.02	1275.	950.	798.	704.	633.	576.	527.	484.	376.
0.10	0.05	1093.	796.	659.	574.	511.	461.	418.	381.	288.
0.10	0.10	949.	676.	551.	474.	418.	374.	336.	303.	223.
0.10	0.20	796.	550.	439.	373.	324.	286.	254.	227.	161.
0.15	0.01	466.	356.	304.	272.	247.	227.	210.	195.	157.
0.15	0.02	423.	319.	271.	240.	217.	199.	183.	169.	134.
0.15	0.05	365.	269.	225.	198.	178.	162.	148.	136.	105.
0.15	0.10	318.	230.	190.	166.	148.	133.	121.	110.	84.
0.15	0.20	269.	190.	154.	132.	117.	104.	94.	85.	63.
0.20	0.01	254.	195.	168.	150.	137.	127.	118.	110.	89.
0.20	0.02	231.	176.	150.	134.	121.	112.	103.	96.	77.
0.20	0.05	199.	149.	125.	111.	100.	91.	84.	77.	61.
0.20	0.10	174.	128.	106.	93.	84.	76.	69.	64.	49.
0.20	0.20	148.	106.	87.	75.	67.	60.	55.	50.	38.
0.25	0.01	167.	129.	111.	100.	92.	85.	79.	74.	61.
0.25	0.02	152.	116.	100.	89.	81.	75.	70.	65.	53.
0.25	0.05	131.	99.	84.	74.	67.	62.	57.	53.	42.
0.25	0.10	115.	85.	71.	63.	57.	52.	47.	44.	34.
0.25	0.20	98.	71.	59.	51.	46.	41.	38.	34.	27.
0.30	0.01	121.	94.	81.	74.	68.	63.	58.	55.	45.
0.30	0.02	110.	85.	73.	65.	60.	55.	52.	48.	39.
0.30	0.05	95.	72.	61.	55.	50.	46.	42.	39.	32.
0.30	0.10	83.	62.	52.	46.	42.	38.	35.	33.	26.
0.30	0.20	71.	52.	43.	38.	34.	31.	28.	26.	20.
0.35	0.01	92.	73.	63.	57.	53.	49.	46.	43.	36.
0.35	0.02	84.	65.	56.	51.	47.	43.	40.	38.	31.
0.35	0.05	73.	56.	48.	43.	39.	36.	33.	31.	25.
0.35	0.10	64.	48.	41.	36.	33.	30.	28.	26.	21.
0.35	0.20	55.	40.	34.	30.	27.	24.	22.	21.	16.
0.40	0.01	74.	58.	51.	46.	42.	40.	37.	35.	29.
0.40	0.02	67.	52.	45.	41.	38.	35.	33.	31.	26.
0.40	0.05	58.	45.	38.	34.	32.	29.	27.	25.	21.
0.40	0.10	51.	39.	33.	29.	27.	25.	23.	21.	17.
0.40	0.20	43.	32.	27.	24.	22.	20.	18.	17.	14.
0.45	0.01	60.	48.	42.	38.	35.	33.	31.	29.	25.
0.45	0.02	55.	43.	37.	34.	31.	29.	27.	26.	22.
0.45	0.05	47.	37.	32.	29.	26.	24.	23.	21.	18.
0.45	0.10	42.	32.	27.	24.	22.	21.	19.	18.	15.
0.45	0.20	36.	27.	23.	20.	18.	17.	16.	14.	12.
0.50	0.01	50.	40.	35.	32.	30.	28.	26.	25.	21.
0.50	0.02	45.	36.	31.	29.	27.	25.	23.	22.	18.
0.50	0.05	39.	31.	27.	24.	22.	21.	19.	18.	15.
0.50	0.10	35.	27.	23.	21.	19.	18.	16.	15.	13.
0.50	0.20	30.	22.	19.	17.	16.	14.	13.	12.	10.
0.55	0.01	42.	34.	30.	27.	25.	24.	23.	21.	18.
0.55	0.02	38.	31.	27.	24.	23.	21.	20.	19.	16.
0.55	0.05	33.	26.	23.	21.	19.	18.	17.	16.	13.
0.55	0.10	29.	23.	20.	18.	16.	15.	14.	13.	11.
0.55	0.20	25.	19.	16.	15.	13.	12.	12.	11.	9.
0.60	0.01	36.	29.	26.	24.	22.	21.	20.	19.	16.

176

TABLE A.3. SAMPLE SIZES PER GROUP FOR TWO-TAILED TEST ON PROPORTIONS
P1 = 0.05

P2	ALPHA	POWER 0.99	0.95	0.90	0.85	0.80	0.75	0.70	0.65	0.50
0.60	0.02	33.	26.	23.	21.	20.	19.	17.	17.	14.
0.60	0.05	28.	22.	20.	18.	17.	16.	15.	14.	12.
0.60	0.10	25.	19.	17.	15.	14.	13.	12.	12.	10.
0.60	0.20	21.	16.	14.	13.	12.	11.	10.	10.	8.
0.65	0.01	31.	25.	22.	21.	19.	18.	17.	16.	14.
0.65	0.02	28.	23.	20.	18.	17.	16.	15.	15.	13.
0.65	0.05	24.	19.	17.	16.	15.	14.	13.	12.	10.
0.65	0.10	21.	17.	15.	13.	12.	12.	11.	10.	9.
0.65	0.20	18.	14.	12.	11.	10.	10.	9.	9.	7.
0.70	0.01	26.	22.	19.	18.	17.	16.	15.	15.	13.
0.70	0.02	24.	20.	18.	16.	15.	14.	14.	13.	11.
0.70	0.05	21.	17.	15.	14.	13.	12.	11.	11.	9.
0.70	0.10	18.	15.	13.	12.	11.	10.	10.	9.	8.
0.70	0.20	16.	12.	11.	10.	9.	9.	8.	8.	7.
0.75	0.01	23.	19.	17.	16.	15.	14.	14.	13.	12.
0.75	0.02	21.	17.	15.	14.	13.	13.	12.	12.	10.
0.75	0.05	18.	15.	13.	12.	11.	11.	10.	10.	9.
0.75	0.10	16.	13.	11.	10.	10.	9.	9.	8.	7.
0.75	0.20	13.	11.	10.	9.	8.	8.	7.	7.	6.
0.80	0.01	20.	16.	15.	14.	13.	13.	12.	12.	10.
0.80	0.02	18.	15.	13.	13.	12.	11.	11.	10.	9.
0.80	0.05	15.	13.	12.	11.	10.	10.	9.	9.	8.
0.80	0.10	13.	11.	10.	9.	9.	8.	8.	8.	7.
0.80	0.20	12.	9.	8.	8.	7.	7.	7.	6.	5.
0.85	0.01	17.	14.	13.	12.	12.	11.	11.	10.	9.
0.85	0.02	15.	13.	12.	11.	11.	10.	10.	9.	8.
0.85	0.05	13.	11.	10.	9.	9.	9.	8.	8.	7.
0.85	0.10	12.	10.	9.	8.	8.	7.	7.	7.	6.
0.85	0.20	10.	8.	7.	7.	6.	6.	6.	6.	5.
0.90	0.01	14.	12.	12.	11.	10.	10.	10.	9.	9.
0.90	0.02	13.	11.	10.	10.	9.	9.	9.	8.	8.
0.90	0.05	11.	10.	9.	8.	8.	8.	7.	7.	7.
0.90	0.10	10.	8.	8.	7.	7.	7.	6.	6.	6.
0.90	0.20	8.	7.	6.	6.	6.	6.	5.	5.	5.
0.95	0.01	12.	11.	10.	10.	9.	9.	9.	9.	8.
0.95	0.02	11.	10.	9.	9.	8.	8.	8.	8.	7.
0.95	0.05	9.	8.	8.	7.	7.	7.	7.	6.	6.
0.95	0.10	8.	7.	7.	6.	6.	6.	6.	6.	5.
0.95	0.20	7.	6.	6.	5.	5.	5.	5.	5.	4.

P1 = 0.10

P2	ALPHA	0.99	0.95	0.90	0.85	POWER 0.80	0.75	0.70	0.65	0.50
0.15	0.01	2176.	1634.	1378.	1219.	1099.	1002.	918.	845.	658.
0.15	0.02	1968.	1456.	1215.	1066.	955.	865.	788.	721.	551.
0.15	0.05	1682.	1213.	996.	862.	764.	684.	617.	558.	412.
0.15	0.10	1454.	1023.	826.	705.	617.	547.	488.	437.	312.
0.15	0.20	1213.	825.	651.	546.	470.	410.	360.	318.	216.
0.20	0.01	646.	490.	416.	370.	335.	307.	283.	261.	207.
0.20	0.02	586.	438.	369.	325.	293.	267.	245.	225.	176.
0.20	0.05	503.	367.	305.	266.	237.	214.	195.	178.	135.
0.20	0.10	437.	312.	255.	220.	195.	174.	157.	142.	105.
0.20	0.20	367.	254.	204.	174.	151.	134.	119.	107.	77.
0.25	0.01	329.	251.	214.	191.	174.	160.	148.	137.	110.
0.25	0.02	298.	225.	190.	169.	153.	140.	129.	119.	94.
0.25	0.05	257.	190.	158.	139.	125.	113.	104.	95.	74.
0.25	0.10	224.	162.	133.	116.	103.	93.	84.	77.	58.
0.25	0.20	189.	133.	108.	92.	81.	73.	65.	59.	44.
0.30	0.01	206.	158.	136.	122.	111.	102.	95.	88.	72.
0.30	0.02	187.	142.	121.	108.	98.	90.	83.	77.	62.
0.30	0.05	161.	120.	101.	89.	80.	73.	67.	62.	49.
0.30	0.10	141.	103.	85.	75.	67.	61.	55.	51.	39.
0.30	0.20	119.	85.	69.	60.	53.	48.	43.	39.	30.
0.35	0.01	144.	111.	96.	86.	79.	73.	68.	63.	52.
0.35	0.02	131.	100.	86.	76.	70.	64.	60.	55.	45.
0.35	0.05	113.	85.	72.	64.	57.	53.	48.	45.	36.
0.35	0.10	99.	73.	61.	54.	48.	44.	40.	37.	29.
0.35	0.20	84.	60.	50.	43.	39.	35.	32.	29.	22.
0.40	0.01	107.	83.	72.	65.	60.	55.	52.	48.	40.
0.40	0.02	98.	75.	65.	58.	53.	49.	45.	42.	35.
0.40	0.05	84.	64.	54.	48.	44.	40.	37.	34.	28.
0.40	0.10	74.	55.	46.	41.	37.	34.	31.	29.	23.
0.40	0.20	63.	46.	38.	33.	30.	27.	25.	23.	18.
0.45	0.01	83.	65.	57.	51.	47.	44.	41.	39.	32.
0.45	0.02	76.	59.	51.	46.	42.	39.	36.	34.	28.
0.45	0.05	66.	50.	43.	38.	35.	32.	30.	28.	22.
0.45	0.10	57.	43.	36.	32.	29.	27.	25.	23.	18.
0.45	0.20	49.	36.	30.	26.	24.	22.	20.	18.	15.
0.50	0.01	67.	53.	46.	42.	39.	36.	34.	32.	26.
0.50	0.02	61.	48.	41.	37.	34.	32.	30.	28.	23.
0.50	0.05	53.	40.	35.	31.	29.	26.	25.	23.	19.
0.50	0.10	46.	35.	30.	26.	24.	22.	21.	19.	15.
0.50	0.20	39.	29.	25.	22.	20.	18.	17.	15.	12.
0.55	0.01	55.	43.	38.	35.	32.	30.	28.	27.	22.
0.55	0.02	50.	39.	34.	31.	29.	27.	25.	23.	20.
0.55	0.05	43.	33.	29.	26.	24.	22.	21.	19.	16.
0.55	0.10	38.	29.	25.	22.	20.	19.	17.	16.	13.
0.55	0.20	32.	24.	20.	18.	17.	15.	14.	13.	11.
0.60	0.01	45.	36.	32.	29.	27.	25.	24.	23.	19.
0.60	0.02	41.	33.	29.	26.	24.	23.	21.	20.	17.
0.60	0.05	36.	28.	24.	22.	20.	19.	18.	17.	14.
0.60	0.10	31.	24.	21.	19.	17.	16.	15.	14.	12.
0.60	0.20	27.	20.	17.	15.	14.	13.	12.	11.	9.
0.65	0.01	38.	31.	27.	25.	23.	22.	21.	20.	17.

TABLE A.3. SAMPLE SIZES PER GROUP FOR TWO-TAILED TEST ON PROPORTIONS
P1 = 0.10

P2	ALPHA	POWER 0.99	0.95	0.90	0.85	0.80	0.75	0.70	0.65	0.50
0.65	0.02	35.	28.	24.	22.	21.	19.	18.	17.	15.
0.65	0.05	30.	24.	21.	19.	17.	16.	15.	14.	12.
0.65	0.10	26.	21.	18.	16.	15.	14.	13.	12.	10.
0.65	0.20	23.	17.	15.	13.	12.	11.	11.	10.	8.
0.70	0.01	32.	26.	23.	22.	20.	19.	18.	17.	15.
0.70	0.02	29.	24.	21.	19.	18.	17.	16.	15.	13.
0.70	0.05	25.	20.	18.	16.	15.	14.	13.	13.	11.
0.70	0.10	22.	18.	15.	14.	13.	12.	11.	11.	9.
0.70	0.20	19.	15.	13.	12.	11.	10.	9.	9.	7.
0.75	0.01	27.	23.	20.	19.	18.	17.	16.	15.	13.
0.75	0.02	25.	20.	18.	17.	16.	15.	14.	13.	12.
0.75	0.05	22.	17.	15.	14.	13.	12.	12.	11.	10.
0.75	0.10	19.	15.	13.	12.	11.	11.	10.	9.	8.
0.75	0.20	16.	13.	11.	10.	9.	9.	8.	8.	7.
0.80	0.01	23.	19.	18.	16.	15.	15.	14.	13.	12.
0.80	0.02	21.	18.	16.	15.	14.	13.	12.	12.	10.
0.80	0.05	18.	15.	13.	12.	12.	11.	10.	10.	9.
0.80	0.10	16.	13.	12.	11.	10.	9.	9.	8.	7.
0.80	0.20	14.	11.	10.	9.	8.	8.	7.	7.	6.
0.85	0.01	20.	17.	15.	14.	14.	13.	12.	12.	11.
0.85	0.02	18.	15.	14.	13.	12.	12.	11.	11.	9.
0.85	0.05	16.	13.	12.	11.	10.	10.	9.	9.	8.
0.85	0.10	14.	11.	10.	9.	9.	8.	8.	8.	7.
0.85	0.20	12.	10.	9.	8.	7.	7.	7.	6.	5.
0.90	0.01	17.	14.	13.	12.	12.	11.	11.	11.	10.
0.90	0.02	15.	13.	12.	11.	11.	10.	10.	9.	8.
0.90	0.05	13.	11.	10.	10.	9.	9.	8.	8.	7.
0.90	0.10	12.	10.	9.	8.	8.	7.	7.	7.	6.
0.90	0.20	10.	8.	7.	7.	7.	6.	6.	6.	5.
0.95	0.01	14.	12.	12.	11.	10.	10.	10.	9.	9.
0.95	0.02	13.	11.	10.	10.	9.	9.	9.	8.	8.
0.95	0.05	11.	10.	9.	8.	8.	8.	7.	7.	7.
0.95	0.10	10.	8.	8.	7.	7.	7.	6.	6.	6.
0.95	0.20	8.	7.	6.	6.	6.	6.	5.	5.	5.

P2	ALPHA	0.99	0.95	0.90	0.85	POWER 0.80	0.75	0.70	0.65	0.50
0.20	0.01	2849.	2133.	1795.	1584.	1426.	1298.	1188.	1091.	844.
0.20	0.02	2574.	1897.	1580.	1383.	1236.	1118.	1016.	937.	703.
0.20	0.05	2196.	1577.	1290.	1114.	984.	879.	790.	713.	521.
0.20	0.10	1896.	1326.	1066.	907.	791.	698.	620.	553.	388.
0.20	0.20	1578.	1065.	835.	697.	597.	518.	453.	397.	264.
0.25	0.01	803.	606.	513.	455.	411.	376.	346.	319.	251.
0.25	0.02	727.	541.	453.	399.	359.	326.	298.	273.	211.
0.25	0.05	622.	452.	373.	325.	289.	260.	235.	214.	160.
0.25	0.10	539.	383.	311.	267.	235.	209.	188.	169.	123.
0.25	0.20	451.	310.	247.	209.	181.	159.	141.	125.	88.
0.30	0.01	393.	299.	254.	226.	205.	188.	174.	161.	128.
0.30	0.02	356.	267.	225.	199.	180.	164.	151.	139.	109.
0.30	0.05	306.	224.	186.	163.	146.	132.	120.	110.	84.
0.30	0.10	266.	191.	156.	135.	120.	108.	97.	88.	66.
0.30	0.20	223.	156.	125.	107.	94.	83.	74.	67.	48.
0.35	0.01	239.	183.	156.	140.	127.	117.	108.	101.	81.
0.35	0.02	217.	164.	139.	123.	112.	102.	94.	87.	69.
0.35	0.05	187.	138.	116.	102.	91.	83.	76.	70.	54.
0.35	0.10	163.	118.	97.	85.	76.	68.	62.	56.	43.
0.35	0.20	137.	97.	79.	68.	60.	53.	48.	43.	32.
0.40	0.01	163.	126.	108.	97.	88.	82.	76.	70.	57.
0.40	0.02	148.	113.	96.	86.	78.	72.	66.	61.	49.
0.40	0.05	128.	95.	80.	71.	64.	58.	54.	49.	39.
0.40	0.10	111.	82.	68.	59.	53.	48.	44.	40.	31.
0.40	0.20	94.	67.	55.	48.	42.	38.	34.	31.	24.
0.45	0.01	119.	92.	80.	72.	66.	61.	57.	53.	43.
0.45	0.02	108.	83.	71.	64.	58.	53.	50.	46.	37.
0.45	0.05	94.	70.	60.	53.	48.	44.	40.	37.	30.
0.45	0.10	82.	60.	51.	44.	40.	36.	33.	31.	24.
0.45	0.20	69.	50.	41.	36.	32.	29.	26.	24.	19.
0.50	0.01	91.	71.	62.	56.	51.	47.	44.	41.	34.
0.50	0.02	83.	64.	55.	49.	45.	42.	39.	36.	30.
0.50	0.05	72.	54.	46.	41.	37.	34.	32.	30.	24.
0.50	0.10	63.	47.	39.	35.	31.	29.	26.	24.	19.
0.50	0.20	53.	39.	32.	28.	25.	23.	21.	19.	15.
0.55	0.01	72.	57.	49.	45.	41.	38.	36.	34.	28.
0.55	0.02	66.	51.	44.	40.	36.	34.	32.	30.	24.
0.55	0.05	57.	43.	37.	33.	30.	28.	26.	24.	20.
0.55	0.10	50.	37.	32.	28.	26.	23.	22.	20.	16.
0.55	0.20	42.	31.	26.	23.	21.	19.	17.	16.	13.
0.60	0.01	58.	46.	40.	37.	34.	32.	30.	28.	23.
0.60	0.02	53.	42.	36.	33.	30.	28.	26.	25.	20.
0.60	0.05	46.	35.	30.	27.	25.	23.	22.	20.	17.
0.60	0.10	40.	31.	26.	23.	21.	20.	18.	17.	14.
0.60	0.20	34.	25.	21.	19.	17.	16.	15.	14.	11.
0.65	0.01	48.	38.	33.	31.	28.	27.	25.	24.	20.
0.65	0.02	44.	34.	30.	27.	25.	24.	22.	21.	17.
0.65	0.05	38.	29.	25.	23.	21.	20.	18.	17.	14.
0.65	0.10	33.	25.	22.	20.	18.	17.	15.	14.	12.
0.65	0.20	28.	21.	18.	16.	15.	13.	12.	12.	9.
0.70	0.01	40.	32.	28.	26.	24.	23.	21.	20.	17.

TABLE A.3. SAMPLE SIZES PER GROUP FOR TWO-TAILED TEST ON PROPORTIONS
P1 = 0.15

P2	ALPHA	0.99	0.95	0.90	0.85	POWER 0.80	0.75	0.70	0.65	0.50
0.70	0.02	36.	29.	25.	23.	21.	20.	19.	18.	15.
0.70	0.05	31.	25.	21.	19.	18.	17.	16.	15.	12.
0.70	0.10	27.	21.	18.	17.	15.	14.	13.	12.	10.
0.70	0.20	23.	18.	15.	14.	13.	12.	11.	10.	8.
0.75	0.01	33.	27.	24.	22.	21.	19.	18.	17.	15.
0.75	0.02	30.	24.	22.	20.	18.	17.	16.	15.	13.
0.75	0.05	26.	21.	18.	17.	15.	14.	14.	13.	11.
0.75	0.10	23.	18.	16.	14.	13.	12.	12.	11.	9.
0.75	0.20	20.	15.	13.	12.	11.	10.	9.	9.	7.
0.80	0.01	28.	23.	21.	19.	18.	17.	16.	15.	13.
0.80	0.02	26.	21.	18.	17.	16.	15.	14.	14.	12.
0.80	0.05	22.	18.	16.	14.	13.	13.	12.	11.	10.
0.80	0.10	19.	15.	14.	12.	11.	11.	10.	10.	8.
0.80	0.20	17.	13.	11.	10.	10.	9.	8.	8.	7.
0.85	0.01	24.	20.	18.	16.	15.	15.	14.	13.	12.
0.85	0.02	22.	18.	16.	15.	14.	13.	13.	12.	10.
0.85	0.05	19.	15.	14.	13.	12.	11.	10.	10.	9.
0.85	0.10	16.	13.	12.	11.	10.	9.	9.	9.	7.
0.85	0.20	14.	11.	10.	9.	8.	8.	7.	7.	6.
0.90	0.01	20.	17.	15.	14.	14.	13.	12.	12.	11.
0.90	0.02	18.	15.	14.	13.	12.	12.	11.	11.	9.
0.90	0.05	16.	13.	12.	11.	10.	10.	9.	9.	8.
0.90	0.10	14.	11.	10.	9.	9.	8.	8.	8.	7.
0.90	0.20	12.	10.	9.	8.	7.	7.	7.	6.	5.
0.95	0.01	17.	14.	13.	12.	12.	11.	11.	10.	9.
0.95	0.02	15.	13.	12.	11.	11.	10.	10.	9.	8.
0.95	0.05	13.	11.	10.	9.	9.	9.	8.	8.	7.
0.95	0.10	12.	10.	9.	8.	8.	7.	7.	7.	6.
0.95	0.20	10.	8.	7.	7.	6.	6.	6.	6.	5.

TABLE A.3. SAMPLE SIZES PER GROUP FOR TWO-TAILED TEST ON PROPORTIONS
P1 = 0.20

P2	ALPHA	\ \ POWER 0.99	0.95	0.90	0.85	0.80	0.75	0.70	0.65	0.50
0.25	0.01	3426.	2561.	2152.	1897.	1707.	1552.	1419.	1301.	1004.
0.25	0.02	3094.	2276.	1893.	1655.	1478.	1334.	1212.	1104.	833.
0.25	0.05	2637.	1889.	1543.	1330.	1172.	1046.	939.	845.	613.
0.25	0.10	2275.	1586.	1272.	1080.	940.	828.	734.	652.	454.
0.25	0.20	1891.	1271.	993.	826.	706.	610.	532.	464.	304.
0.30	0.01	935.	704.	595.	527.	476.	434.	399.	367.	287.
0.30	0.02	846.	627.	525.	461.	414.	376.	343.	314.	241.
0.30	0.05	723.	524.	431.	374.	332.	298.	269.	244.	182.
0.30	0.10	626.	442.	358.	307.	269.	239.	214.	192.	139.
0.30	0.20	523.	358.	283.	239.	206.	181.	159.	141.	97.
0.35	0.01	446.	338.	287.	255.	231.	212.	195.	180.	143.
0.35	0.02	404.	302.	254.	225.	202.	184.	169.	155.	121.
0.35	0.05	347.	253.	210.	183.	163.	148.	134.	122.	93.
0.35	0.10	301.	215.	176.	152.	134.	120.	108.	97.	72.
0.35	0.20	252.	175.	140.	119.	104.	92.	82.	73.	52.
0.40	0.01	266.	203.	173.	154.	140.	129.	119.	111.	89.
0.40	0.02	241.	182.	154.	136.	123.	113.	104.	96.	75.
0.40	0.05	207.	153.	127.	112.	100.	91.	83.	76.	59.
0.40	0.10	180.	130.	107.	93.	83.	74.	67.	61.	46.
0.40	0.20	152.	107.	86.	74.	65.	58.	52.	47.	34.
0.45	0.01	179.	137.	117.	105.	96.	88.	82.	76.	62.
0.45	0.02	162.	123.	104.	93.	84.	77.	71.	66.	53.
0.45	0.05	139.	104.	87.	77.	69.	63.	58.	53.	41.
0.45	0.10	122.	89.	73.	64.	57.	52.	47.	43.	33.
0.45	0.20	103.	73.	59.	51.	45.	41.	37.	33.	25.
0.50	0.01	129.	99.	86.	77.	70.	65.	60.	56.	46.
0.50	0.02	117.	89.	76.	68.	62.	57.	53.	49.	40.
0.50	0.05	101.	75.	64.	56.	51.	47.	43.	40.	31.
0.50	0.10	88.	65.	54.	47.	42.	39.	35.	32.	25.
0.50	0.20	74.	53.	44.	38.	34.	31.	28.	25.	19.
0.55	0.01	97.	76.	65.	59.	54.	50.	47.	44.	36.
0.55	0.02	88.	68.	58.	52.	48.	44.	41.	38.	31.
0.55	0.05	76.	58.	49.	43.	39.	36.	33.	31.	25.
0.55	0.10	67.	49.	42.	37.	33.	30.	28.	26.	20.
0.55	0.20	56.	41.	34.	30.	27.	24.	22.	20.	16.
0.60	0.01	76.	59.	52.	47.	43.	40.	37.	35.	29.
0.60	0.02	69.	53.	46.	42.	38.	35.	33.	31.	25.
0.60	0.05	60.	45.	39.	35.	32.	29.	27.	25.	20.
0.60	0.10	52.	39.	33.	29.	27.	24.	22.	21.	17.
0.60	0.20	44.	32.	27.	24.	21.	20.	18.	16.	13.
0.65	0.01	61.	48.	42.	38.	35.	33.	31.	29.	24.
0.65	0.02	55.	43.	37.	34.	31.	29.	27.	25.	21.
0.65	0.05	48.	37.	31.	28.	26.	24.	22.	21.	17.
0.65	0.10	42.	32.	27.	24.	22.	20.	19.	17.	14.
0.65	0.20	35.	26.	22.	20.	18.	16.	15.	14.	11.
0.70	0.01	49.	39.	34.	31.	29.	27.	26.	24.	20.
0.70	0.02	45.	35.	31.	28.	26.	24.	23.	21.	18.
0.70	0.05	39.	30.	26.	23.	22.	20.	19.	17.	15.
0.70	0.10	34.	26.	22.	20.	18.	17.	16.	15.	12.
0.70	0.20	29.	22.	18.	16.	15.	14.	13.	12.	10.
0.75	0.01	41.	33.	29.	26.	24.	23.	22.	20.	17.

182

TABLE A.3. SAMPLE SIZES PER GROUP FOR TWO-TAILED TEST ON PROPORTIONS
P1 = 0.20

P2	ALPHA	POWER 0.99	0.95	0.90	0.85	0.80	0.75	0.70	0.65	0.50
0.75	0.02	37.	29.	26.	24.	22.	20.	19.	18.	15.
0.75	0.05	32.	25.	22.	20.	18.	17.	16.	15.	13.
0.75	0.10	28.	22.	19.	17.	16.	14.	13.	13.	10.
0.75	0.20	24.	18.	16.	14.	13.	12.	11.	10.	8.
0.80	0.01	34.	27.	24.	22.	21.	20.	19.	18.	15.
0.80	0.02	31.	25.	22.	20.	19.	17.	16.	16.	13.
0.80	0.05	27.	21.	18.	17.	16.	15.	14.	13.	11.
0.80	0.10	23.	18.	16.	14.	13.	12.	12.	11.	9.
0.80	0.20	20.	15.	13.	12.	11.	10.	10.	9.	7.
0.85	0.01	28.	23.	21.	19.	18.	17.	16.	15.	13.
0.85	0.02	26.	21.	18.	17.	16.	15.	14.	14.	12.
0.85	0.05	22.	18.	16.	14.	13.	13.	12.	11.	10.
0.85	0.10	19.	15.	14.	12.	11.	11.	10.	10.	8.
0.85	0.20	17.	13.	11.	10.	10.	9.	8.	8.	7.
0.90	0.01	23.	19.	18.	16.	15.	15.	14.	13.	12.
0.90	0.02	21.	18.	16.	15.	14.	13.	12.	12.	10.
0.90	0.05	18.	15.	13.	12.	12.	11.	10.	10.	9.
0.90	0.10	16.	13.	12.	11.	10.	9.	9.	8.	7.
0.90	0.20	14.	11.	10.	9.	8.	8.	7.	7.	6.
0.95	0.01	20.	16.	15.	14.	13.	13.	12.	12.	10.
0.95	0.02	18.	15.	13.	13.	12.	11.	11.	10.	9.
0.95	0.05	15.	13.	12.	11.	10.	10.	9.	9.	8.
0.95	0.10	13.	11.	10.	9.	9.	8.	8.	8.	7.
0.95	0.20	12.	9.	8.	8.	7.	7.	7.	6.	5.

TABLE A.3. SAMPLE SIZES PER GROUP FOR TWO-TAILED TEST ON PROPORTIONS
P1 = 0.25

P2	ALPHA	0.99	0.95	0.90	0.85	POWER 0.80	0.75	0.70	0.65	0.50
0.30	0.01	3907.	2917.	2450.	2159.	1940.	1763.	1611.	1477.	1137.
0.30	0.02	3527.	2591.	2153.	1881.	1678.	1514.	1374.	1251.	941.
0.30	0.05	3005.	2149.	1753.	1509.	1330.	1185.	1062.	955.	690.
0.30	0.10	2590.	1803.	1443.	1224.	1064.	936.	828.	735.	508.
0.30	0.20	2151.	1443.	1125.	934.	796.	687.	597.	520.	337.
0.35	0.01	1043.	784.	662.	585.	528.	482.	442.	407.	317.
0.35	0.02	943.	699.	584.	512.	459.	416.	380.	347.	266.
0.35	0.05	806.	582.	478.	415.	367.	329.	297.	269.	199.
0.35	0.10	697.	491.	397.	339.	297.	264.	235.	211.	151.
0.35	0.20	582.	396.	313.	263.	227.	198.	174.	154.	105.
0.40	0.01	489.	370.	314.	279.	252.	231.	212.	196.	155.
0.40	0.02	443.	330.	278.	245.	220.	200.	183.	168.	131.
0.40	0.05	379.	276.	229.	199.	178.	160.	145.	132.	100.
0.40	0.10	329.	234.	191.	164.	145.	129.	116.	105.	77.
0.40	0.20	276.	190.	152.	129.	112.	99.	88.	78.	55.
0.45	0.01	287.	219.	186.	166.	151.	138.	128.	118.	94.
0.45	0.02	260.	196.	165.	146.	132.	121.	111.	102.	80.
0.45	0.05	224.	164.	137.	120.	107.	97.	88.	81.	62.
0.45	0.10	194.	140.	115.	99.	88.	79.	72.	65.	49.
0.45	0.20	163.	114.	92.	79.	69.	61.	55.	49.	36.
0.50	0.01	190.	146.	125.	111.	102.	93.	87.	80.	65.
0.50	0.02	173.	130.	111.	98.	89.	82.	75.	70.	55.
0.50	0.05	148.	110.	92.	81.	73.	66.	61.	56.	43.
0.50	0.10	129.	94.	78.	68.	60.	54.	49.	45.	34.
0.50	0.20	109.	77.	63.	54.	48.	43.	38.	35.	26.
0.55	0.01	135.	104.	90.	80.	74.	68.	63.	59.	48.
0.55	0.02	123.	94.	80.	71.	65.	60.	55.	51.	41.
0.55	0.05	106.	79.	67.	59.	53.	49.	45.	41.	32.
0.55	0.10	92.	68.	56.	49.	44.	40.	37.	34.	26.
0.55	0.20	78.	56.	46.	40.	35.	32.	29.	26.	20.
0.60	0.01	101.	79.	68.	61.	56.	52.	48.	45.	37.
0.60	0.02	92.	71.	60.	54.	49.	46.	42.	39.	32.
0.60	0.05	79.	60.	51.	45.	41.	37.	34.	32.	25.
0.60	0.10	69.	51.	43.	38.	34.	31.	28.	26.	21.
0.60	0.20	59.	42.	35.	31.	27.	25.	23.	21.	16.
0.65	0.01	78.	61.	53.	48.	44.	41.	38.	36.	30.
0.65	0.02	71.	55.	47.	43.	39.	36.	34.	31.	26.
0.65	0.05	61.	47.	40.	35.	32.	30.	28.	26.	21.
0.65	0.10	54.	40.	34.	30.	27.	25.	23.	21.	17.
0.65	0.20	45.	33.	28.	24.	22.	20.	18.	17.	13.
0.70	0.01	62.	49.	43.	39.	36.	33.	31.	29.	24.
0.70	0.02	56.	44.	38.	34.	32.	29.	27.	26.	21.
0.70	0.05	49.	37.	32.	29.	26.	24.	23.	21.	17.
0.70	0.10	43.	32.	27.	24.	22.	20.	19.	18.	14.
0.70	0.20	36.	27.	23.	20.	18.	16.	15.	14.	11.
0.75	0.01	50.	40.	35.	32.	29.	27.	26.	24.	20.
0.75	0.02	45.	36.	31.	28.	26.	24.	23.	21.	18.
0.75	0.05	39.	30.	26.	24.	22.	20.	19.	18.	15.
0.75	0.10	34.	26.	22.	20.	18.	17.	16.	15.	12.
0.75	0.20	29.	22.	19.	17.	15.	14.	13.	12.	10.
0.80	0.01	41.	33.	29.	26.	24.	23.	22.	20.	17.

TABLE A.3. SAMPLE SIZES PER GROUP FOR TWO-TAILED TEST ON PROPORTIONS
P1 = 0.25

P2	ALPHA	POWER								
		0.99	0.95	0.90	0.85	0.80	0.75	0.70	0.65	0.50
0.80	0.02	37.	29.	26.	24.	22.	20.	19.	18.	15.
0.80	0.05	32.	25.	22.	20.	18.	17.	16.	15.	13.
0.80	0.10	28.	22.	19.	17.	16.	14.	13.	13.	10.
0.80	0.20	24.	18.	16.	14.	13.	12.	11.	10.	8.
0.85	0.01	33.	27.	24.	22.	21.	19.	18.	17.	15.
0.85	0.02	30.	24.	22.	20.	18.	17.	16.	15.	13.
0.85	0.05	26.	21.	18.	17.	15.	14.	14.	13.	11.
0.85	0.10	23.	18.	16.	14.	13.	12.	12.	11.	9.
0.85	0.20	20.	15.	13.	12.	11.	10.	9.	9.	7.
0.90	0.01	27.	23.	20.	19.	18.	17.	16.	15.	13.
0.90	0.02	25.	20.	18.	17.	16.	15.	14.	13.	12.
0.90	0.05	22.	17.	15.	14.	13.	12.	12.	11.	10.
0.90	0.10	19.	15.	13.	12.	11.	11.	10.	9.	8.
0.90	0.20	16.	13.	11.	10.	9.	9.	8.	8.	7.
0.95	0.01	23.	19.	17.	16.	15.	14.	14.	13.	12.
0.95	0.02	21.	17.	15.	14.	13.	13.	12.	12.	10.
0.95	0.05	18.	15.	13.	12.	11.	11.	10.	10.	9.
0.95	0.10	16.	13.	11.	10.	10.	9.	9.	8.	7.
0.95	0.20	13.	11.	10.	9.	8.	8.	7.	7.	6.

TABLE A.3. SAMPLE SIZES PER GROUP FOR TWO-TAILED TEST ON PROPORTIONS
P1 = 0.30

P2	ALPHA	POWER								
		0.99	0.95	0.90	0.85	0.80	0.75	0.70	0.65	0.50
0.35	0.01	4291.	3202.	2688.	2367.	2127.	1932.	1765.	1617.	1243.
0.35	0.02	3873.	2844.	2361.	2062.	1839.	1658.	1504.	1369.	1028.
0.35	0.05	3299.	2357.	1921.	1653.	1455.	1296.	1161.	1043.	752.
0.35	0.10	2843.	1976.	1580.	1339.	1163.	1022.	904.	801.	552.
0.35	0.20	2360.	1580.	1230.	1020.	868.	749.	649.	564.	364.
0.40	0.01	1127.	847.	714.	631.	569.	519.	476.	437.	341.
0.40	0.02	1019.	754.	629.	552.	494.	448.	408.	373.	285.
0.40	0.05	870.	628.	515.	446.	395.	354.	319.	288.	213.
0.40	0.10	752.	529.	427.	365.	319.	283.	252.	225.	161.
0.40	0.20	627.	426.	336.	282.	243.	212.	186.	164.	111.
0.45	0.01	521.	394.	334.	296.	268.	245.	225.	208.	164.
0.45	0.02	472.	351.	295.	260.	234.	212.	194.	178.	138.
0.45	0.05	404.	294.	243.	211.	188.	169.	153.	140.	105.
0.45	0.10	350.	249.	202.	174.	153.	137.	123.	111.	81.
0.45	0.20	293.	202.	161.	136.	118.	104.	92.	82.	58.
0.50	0.01	302.	230.	196.	174.	158.	145.	134.	124.	99.
0.50	0.02	274.	205.	173.	153.	138.	126.	116.	107.	84.
0.50	0.05	235.	172.	143.	125.	112.	101.	92.	84.	65.
0.50	0.10	204.	146.	120.	104.	92.	83.	75.	68.	50.
0.50	0.20	171.	120.	96.	82.	72.	64.	57.	51.	37.
0.55	0.01	198.	151.	129.	116.	105.	97.	90.	83.	67.
0.55	0.02	179.	136.	115.	102.	92.	85.	78.	72.	57.
0.55	0.05	154.	114.	95.	84.	75.	68.	63.	57.	45.
0.55	0.10	134.	97.	80.	70.	62.	56.	51.	46.	35.
0.55	0.20	113.	80.	65.	56.	49.	44.	39.	36.	26.
0.60	0.01	139.	107.	92.	83.	76.	70.	65.	60.	49.
0.60	0.02	127.	96.	82.	73.	67.	61.	56.	52.	42.
0.60	0.05	109.	81.	68.	60.	54.	50.	46.	42.	33.
0.60	0.10	95.	69.	58.	51.	45.	41.	37.	34.	27.
0.60	0.20	80.	57.	47.	41.	36.	32.	29.	27.	20.
0.65	0.01	103.	80.	69.	62.	57.	53.	49.	46.	38.
0.65	0.02	94.	72.	62.	55.	50.	46.	43.	40.	32.
0.65	0.05	81.	61.	51.	46.	41.	38.	35.	32.	26.
0.65	0.10	70.	52.	44.	38.	35.	31.	29.	27.	21.
0.65	0.20	60.	43.	36.	31.	28.	25.	23.	21.	16.
0.70	0.01	79.	62.	53.	48.	44.	41.	39.	36.	30.
0.70	0.02	72.	55.	48.	43.	39.	36.	34.	32.	26.
0.70	0.05	62.	47.	40.	36.	33.	30.	28.	26.	21.
0.70	0.10	54.	40.	34.	30.	27.	25.	23.	21.	17.
0.70	0.20	46.	34.	28.	25.	22.	20.	18.	17.	13.
0.75	0.01	62.	49.	43.	39.	36.	33.	31.	29.	24.
0.75	0.02	56.	44.	38.	34.	32.	29.	27.	26.	21.
0.75	0.05	49.	37.	32.	29.	26.	24.	23.	21.	17.
0.75	0.10	43.	32.	27.	24.	22.	20.	19.	18.	14.
0.75	0.20	36.	27.	23.	20.	18.	16.	15.	14.	11.
0.80	0.01	49.	39.	34.	31.	29.	27.	26.	24.	20.
0.80	0.02	45.	35.	31.	28.	26.	24.	23.	21.	18.
0.80	0.05	39.	30.	26.	23.	22.	20.	19.	17.	15.
0.80	0.10	34.	26.	22.	20.	18.	17.	16.	15.	12.
0.80	0.20	29.	22.	18.	16.	15.	14.	13.	12.	10.
0.85	0.01	40.	32.	28.	26.	24.	23.	21.	20.	17.
0.85	0.02	36.	29.	25.	23.	21.	20.	19.	18.	15.
0.85	0.05	31.	25.	21.	19.	18.	17.	16.	15.	12.
0.85	0.10	27.	21.	18.	17.	15.	14.	13.	12.	10.
0.85	0.20	23.	18.	15.	14.	13.	12.	11.	10.	8.
0.90	0.01	32.	26.	23.	22.	20.	19.	18.	17.	15.
0.90	0.02	29.	24.	21.	19.	18.	17.	16.	15.	13.
0.90	0.05	25.	20.	18.	16.	15.	14.	13.	13.	11.
0.90	0.10	22.	18.	15.	14.	13.	12.	11.	11.	9.
0.90	0.20	19.	15.	13.	12.	11.	10.	9.	9.	7.
0.95	0.01	26.	22.	19.	18.	17.	16.	15.	15.	13.
0.95	0.02	24.	20.	18.	16.	15.	14.	14.	13.	11.
0.95	0.05	21.	17.	15.	14.	13.	12.	11.	11.	9.
0.95	0.10	18.	15.	13.	12.	11.	10.	10.	9.	8.
0.95	0.20	16.	12.	11.	10.	9.	9.	8.	8.	7.

TABLE A.3. SAMPLE SIZES PER GROUP FOR TWO-TAILED TEST ON PROPORTIONS
P1 = 0.35

P2	ALPHA	POWER								
		0.99	0.95	0.90	0.85	0.80	0.75	0.70	0.65	0.50
0.40	0.01	4580.	3416.	2867.	2524.	2268.	2059.	1880.	1723.	1323.
0.40	0.02	4133.	3033.	2518.	2198.	1960.	1766.	1602.	1457.	1093.
0.40	0.05	3519.	2513.	2047.	1761.	1549.	1379.	1235.	1109.	798.
0.40	0.10	3032.	2106.	1683.	1425.	1237.	1086.	960.	851.	585.
0.40	0.20	2516.	1682.	1309.	1085.	923.	795.	689.	598.	384.
0.45	0.01	1187.	891.	751.	664.	598.	545.	500.	459.	357.
0.45	0.02	1073.	793.	662.	580.	520.	470.	428.	391.	298.
0.45	0.05	916.	660.	542.	469.	415.	371.	334.	302.	223.
0.45	0.10	792.	556.	448.	383.	335.	296.	264.	236.	167.
0.45	0.20	660.	448.	353.	295.	254.	221.	194.	171.	115.
0.50	0.01	543.	410.	347.	308.	278.	254.	234.	216.	170.
0.50	0.02	491.	365.	307.	270.	243.	220.	202.	185.	143.
0.50	0.05	420.	305.	252.	219.	195.	176.	159.	144.	108.
0.50	0.10	364.	258.	210.	180.	159.	141.	127.	114.	83.
0.50	0.20	305.	209.	167.	141.	122.	107.	95.	85.	59.
0.55	0.01	311.	237.	201.	179.	162.	149.	137.	127.	101.
0.55	0.02	282.	211.	178.	158.	142.	130.	119.	110.	86.
0.55	0.05	242.	177.	147.	129.	115.	104.	95.	86.	66.
0.55	0.10	210.	151.	123.	107.	94.	85.	76.	69.	52.
0.55	0.20	176.	123.	99.	84.	74.	65.	58.	52.	38.
0.60	0.01	202.	154.	132.	118.	107.	99.	91.	85.	68.
0.60	0.02	183.	138.	117.	104.	94.	86.	79.	73.	58.
0.60	0.05	157.	116.	97.	85.	77.	70.	64.	58.	45.
0.60	0.10	137.	99.	82.	71.	63.	57.	52.	47.	36.
0.60	0.20	115.	81.	66.	57.	50.	44.	40.	36.	27.
0.65	0.01	141.	108.	93.	83.	76.	70.	65.	61.	49.
0.65	0.02	128.	97.	83.	74.	67.	62.	57.	53.	42.
0.65	0.05	110.	82.	69.	61.	55.	50.	46.	42.	33.
0.65	0.10	96.	70.	58.	51.	46.	41.	38.	35.	27.
0.65	0.20	81.	58.	47.	41.	36.	33.	29.	27.	20.
0.70	0.01	103.	80.	69.	62.	57.	53.	49.	46.	38.
0.70	0.02	94.	72.	62.	55.	50.	46.	43.	40.	32.
0.70	0.05	81.	61.	51.	46.	41.	38.	35.	32.	26.
0.70	0.10	70.	52.	44.	38.	35.	31.	29.	27.	21.
0.70	0.20	60.	43.	36.	31.	28.	25.	23.	21.	16.
0.75	0.01	78.	61.	53.	48.	44.	41.	38.	36.	30.
0.75	0.02	71.	55.	47.	43.	39.	36.	34.	31.	26.
0.75	0.05	61.	47.	40.	35.	32.	30.	28.	26.	21.
0.75	0.10	54.	40.	34.	30.	27.	25.	23.	21.	17.
0.75	0.20	45.	33.	28.	24.	22.	20.	18.	17.	13.
0.80	0.01	61.	48.	42.	38.	35.	33.	31.	29.	24.
0.80	0.02	55.	43.	37.	34.	31.	29.	27.	25.	21.
0.80	0.05	48.	37.	31.	28.	26.	24.	22.	21.	17.
0.80	0.10	42.	32.	27.	24.	22.	20.	19.	17.	14.
0.80	0.20	35.	26.	22.	20.	18.	16.	15.	14.	11.
0.85	0.01	48.	38.	33.	31.	28.	27.	25.	24.	20.
0.85	0.02	44.	34.	30.	27.	25.	24.	22.	21.	17.
0.85	0.05	38.	29.	25.	23.	21.	20.	18.	17.	14.
0.85	0.10	33.	25.	22.	20.	18.	17.	15.	14.	12.
0.85	0.20	28.	21.	18.	16.	15.	13.	12.	12.	9.
0.90	0.01	38.	31.	27.	25.	23.	22.	21.	20.	17.
0.90	0.02	35.	28.	24.	22.	21.	19.	18.	17.	15.
0.90	0.05	30.	24.	21.	19.	17.	16.	15.	14.	12.
0.90	0.10	26.	21.	18.	16.	15.	14.	13.	12.	10.
0.90	0.20	23.	17.	15.	13.	12.	11.	11.	10.	8.
0.95	0.01	31.	25.	22.	21.	19.	18.	17.	16.	14.
0.95	0.02	28.	23.	20.	18.	17.	16.	15.	15.	13.
0.95	0.05	24.	19.	17.	16.	15.	14.	13.	12.	10.
0.95	0.10	21.	17.	15.	13.	12.	12.	11.	10.	9.
0.95	0.20	18.	14.	12.	11.	10.	10.	9.	9.	7.

TABLE A.3. SAMPLE SIZES PER GROUP FOR TWO-TAILED TEST ON PROPORTIONS
P1 = 0.40

P2	ALPHA	0.99	0.95	0.90	0.85	POWER 0.80	0.75	0.70	0.65	0.50
0.45	0.01	4772.	3559.	2986.	2628.	2361.	2144.	1957.	1793.	1376.
0.45	0.02	4306.	3159.	2622.	2288.	2040.	1839.	1667.	1516.	1137.
0.45	0.05	3666.	2617.	2131.	1833.	1612.	1435.	1285.	1153.	829.
0.45	0.10	3158.	2192.	1752.	1483.	1286.	1130.	998.	884.	606.
0.45	0.20	2620.	1751.	1362.	1128.	959.	825.	715.	620.	397.
0.50	0.01	1223.	918.	774.	683.	616.	561.	514.	473.	367.
0.50	0.02	1106.	817.	681.	597.	535.	484.	441.	402.	307.
0.50	0.05	944.	680.	557.	482.	426.	382.	344.	311.	228.
0.50	0.10	816.	572.	461.	393.	344.	304.	271.	242.	172.
0.50	0.20	680.	461.	362.	303.	261.	227.	199.	175.	118.
0.55	0.01	553.	418.	354.	313.	283.	259.	238.	220.	173.
0.55	0.02	501.	372.	312.	275.	247.	224.	205.	188.	145.
0.55	0.05	429.	311.	257.	223.	199.	179.	162.	147.	110.
0.55	0.10	371.	263.	214.	184.	162.	144.	129.	116.	85.
0.55	0.20	310.	213.	170.	143.	124.	109.	97.	86.	60.
0.60	0.01	314.	239.	203.	181.	164.	150.	139.	128.	102.
0.60	0.02	285.	213.	180.	159.	143.	131.	120.	110.	86.
0.60	0.05	244.	179.	149.	130.	116.	105.	95.	87.	67.
0.60	0.10	212.	152.	124.	108.	95.	85.	77.	70.	52.
0.60	0.20	178.	124.	99.	85.	74.	66.	59.	53.	38.
0.65	0.01	202.	154.	132.	118.	107.	99.	91.	85.	68.
0.65	0.02	183.	138.	117.	104.	94.	86.	79.	73.	58.
0.65	0.05	157.	116.	97.	85.	77.	70.	64.	58.	45.
0.65	0.10	137.	99.	82.	71.	63.	57.	52.	47.	36.
0.65	0.20	115.	81.	66.	57.	50.	44.	40.	36.	27.
0.70	0.01	139.	107.	92.	83.	76.	70.	65.	60.	49.
0.70	0.02	127.	96.	82.	73.	67.	61.	56.	52.	42.
0.70	0.05	109.	81.	68.	60.	54.	50.	46.	42.	33.
0.70	0.10	95.	69.	58.	51.	45.	41.	37.	34.	27.
0.70	0.20	80.	57.	47.	41.	36.	32.	29.	27.	20.
0.75	0.01	101.	79.	68.	61.	56.	52.	48.	45.	37.
0.75	0.02	92.	71.	60.	54.	49.	46.	42.	39.	32.
0.75	0.05	79.	60.	51.	45.	41.	37.	34.	32.	25.
0.75	0.10	69.	51.	43.	38.	34.	31.	28.	26.	21.
0.75	0.20	59.	42.	35.	31.	27.	25.	23.	21.	16.
0.80	0.01	76.	59.	52.	47.	43.	40.	37.	35.	29.
0.80	0.02	69.	53.	46.	42.	38.	35.	33.	31.	25.
0.80	0.05	60.	45.	39.	35.	32.	29.	27.	25.	20.
0.80	0.10	52.	39.	33.	29.	27.	24.	22.	21.	17.
0.80	0.20	44.	32.	27.	24.	21.	20.	18.	16.	13.
0.85	0.01	58.	46.	40.	37.	34.	32.	30.	28.	23.
0.85	0.02	53.	42.	36.	33.	30.	28.	26.	25.	20.
0.85	0.05	46.	35.	30.	27.	25.	23.	22.	20.	17.
0.85	0.10	40.	31.	26.	23.	21.	20.	18.	17.	14.
0.85	0.20	34.	25.	21.	19.	17.	16.	15.	14.	11.
0.90	0.01	45.	36.	32.	29.	27.	25.	24.	23.	19.
0.90	0.02	41.	33.	29.	26.	24.	23.	21.	20.	17.
0.90	0.05	36.	28.	24.	22.	20.	19.	18.	17.	14.
0.90	0.10	31.	24.	21.	19.	17.	16.	15.	14.	12.
0.90	0.20	27.	20.	17.	15.	14.	13.	12.	11.	9.
0.95	0.01	36.	29.	26.	24.	22.	21.	20.	19.	16.
0.95	**0.02**	33.	26.	23.	21.	20.	19.	17.	17.	14.
0.95	0.05	28.	22.	20.	18.	17.	16.	15.	14.	12.
0.95	0.10	25.	19.	17.	15.	14.	13.	12.	12.	10.
0.95	0.20	21.	16.	14.	13.	12.	11.	10.	10.	8.

TABLE A.3. SAMPLE SIZES PER GROUP FOR TWO-TAILED TEST ON PROPORTIONS
P1 = 0.45

P2	ALPHA	POWER 0.99	0.95	0.90	0.85	0.80	0.75	0.70	0.65	0.50
0.50	0.01	4868.	3630.	3045.	2681.	2408.	2186.	1996.	1828.	1402.
0.50	0.02	4393.	3222.	2674.	2333.	2080.	1875.	1700.	1545.	1158.
0.50	0.05	3740.	2669.	2174.	1869.	1644.	1463.	1309.	1176.	845.
0.50	0.10	3221.	2236.	1786.	1512.	1311.	1151.	1017.	900.	617.
0.50	0.20	2672.	1785.	1388.	1149.	977.	841.	728.	631.	404.
0.55	0.01	1235.	927.	781.	690.	622.	566.	519.	477.	371.
0.55	0.02	1117.	825.	688.	603.	540.	488.	445.	406.	309.
0.55	0.05	953.	686.	563.	487.	430.	385.	347.	313.	230.
0.55	0.10	823.	578.	465.	397.	347.	307.	273.	244.	173.
0.55	0.20	686.	465.	366.	306.	263.	229.	201.	176.	119.
0.60	0.01	553.	418.	354.	313.	283.	259.	238.	220.	173.
0.60	0.02	501.	372.	312.	275.	247.	224.	205.	188.	145.
0.60	0.05	429.	311.	257.	223.	199.	179.	162.	147.	110.
0.60	0.10	371.	263.	214.	184.	162.	144.	129.	116.	85.
0.60	0.20	310.	213.	170.	143.	124.	109.	97.	86.	60.
0.65	0.01	311.	237.	201.	179.	162.	149.	137.	127.	101.
0.65	0.02	282.	211.	178.	158.	142.	130.	119.	110.	86.
0.65	0.05	242.	177.	147.	129.	115.	104.	95.	86.	66.
0.65	0.10	210.	151.	123.	107.	94.	85.	76.	69.	52.
0.65	0.20	176.	123.	99.	84.	74.	65.	58.	52.	38.
0.70	0.01	198.	151.	129.	116.	105.	97.	90.	83.	67.
0.70	0.02	179.	136.	115.	102.	92.	85.	78.	72.	57.
0.70	0.05	154.	114.	95.	84.	75.	68.	63.	57.	45.
0.70	0.10	134.	97.	80.	70.	62.	56.	51.	46.	35.
0.70	0.20	113.	80.	65.	56.	49.	44.	39.	36.	26.
0.75	0.01	135.	104.	90.	80.	74.	68.	63.	59.	48.
0.75	0.02	123.	94.	80.	71.	65.	60.	55.	51.	41.
0.75	0.05	106.	79.	67.	59.	53.	49.	45.	41.	32.
0.75	0.10	92.	68.	56.	49.	44.	40.	37.	34.	26.
0.75	0.20	78.	56.	46.	40.	35.	32.	29.	26.	20.
0.80	0.01	97.	76.	65.	59.	54.	50.	47.	44.	36.
0.80	0.02	88.	68.	58.	52.	48.	44.	41.	38.	31.
0.80	0.05	76.	58.	49.	43.	39.	36.	33.	31.	25.
0.80	0.10	67.	49.	42.	37.	33.	30.	28.	26.	20.
0.80	0.20	56.	41.	34.	30.	27.	24.	22.	20.	16.
0.85	0.01	72.	57.	49.	45.	41.	38.	36.	34.	28.
0.85	0.02	66.	51.	44.	40.	36.	34.	32.	30.	24.
0.85	0.05	57.	43.	37.	33.	30.	28.	26.	24.	20.
0.85	0.10	50.	37.	32.	28.	26.	23.	22.	20.	16.
0.85	0.20	42.	31.	26.	23.	21.	19.	17.	16.	13.
0.90	0.01	55.	43.	38.	35.	32.	30.	28.	27.	22.
0.90	0.02	50.	39.	34.	31.	29.	27.	25.	23.	20.
0.90	0.05	43.	33.	29.	26.	24.	22.	21.	19.	16.
0.90	0.10	38.	29.	25.	22.	20.	19.	17.	16.	13.
0.90	0.20	32.	24.	20.	18.	17.	15.	14.	13.	11.
0.95	0.01	42.	34.	30.	27.	25.	24.	23.	21.	18.
0.95	0.02	38.	31.	27.	24.	23.	21.	20.	19.	16.
0.95	0.05	33.	26.	23.	21.	19.	18.	17.	16.	13.
0.95	0.10	29.	23.	20.	18.	16.	15.	14.	13.	11.
0.95	0.20	25.	19.	16.	15.	13.	12.	12.	11.	9.

TABLE A.3. SAMPLE SIZES PER GROUP FOR TWO-TAILED TEST ON PROPORTIONS
P1 = 0.50

P2	ALPHA	POWER 0.99	0.95	0.90	0.85	0.80	0.75	0.70	0.65	0.50
0.55	0.01	4868.	3630.	3045.	2681.	2408.	2186.	1996.	1828.	1402.
0.55	0.02	4393.	3222.	2674.	2333.	2080.	1875.	1700.	1545.	1158.
0.55	0.05	3740.	2669.	2174.	1869.	1644.	1463.	1309.	1176.	845.
0.55	0.10	3221.	2236.	1786.	1512.	1311.	1151.	1017.	900.	617.
0.55	0.20	2672.	1785.	1388.	1149.	977.	841.	728.	631.	404.
0.60	0.01	1223.	918.	774.	683.	616.	561.	514.	473.	367.
0.60	0.02	1106.	817.	681.	597.	535.	484.	441.	402.	307.
0.60	0.05	944.	680.	557.	482.	426.	382.	344.	311.	228.
0.60	0.10	816.	572.	461.	393.	344.	304.	271.	242.	172.
0.60	0.20	680.	461.	362.	303.	261.	227.	199.	175.	118.
0.65	0.01	543.	410.	347.	308.	278.	254.	234.	216.	170.
0.65	0.02	491.	365.	307.	270.	243.	220.	202.	185.	143.
0.65	0.05	420.	305.	252.	219.	195.	176.	159.	144.	108.
0.65	0.10	364.	258.	210.	180.	159.	141.	127.	114.	83.
0.65	0.20	305.	209.	167.	141.	122.	107.	95.	85.	59.
0.70	0.01	302.	230.	196.	174.	158.	145.	134.	124.	99.
0.70	0.02	274.	205.	173.	153.	138.	126.	116.	107.	84.
0.70	0.05	235.	172.	143.	125.	112.	101.	92.	84.	65.
0.70	0.10	204.	146.	120.	104.	92.	83.	75.	68.	50.
0.70	0.20	171.	120.	96.	82.	72.	64.	57.	51.	37.
0.75	0.01	190.	146.	125.	111.	102.	93.	87.	80.	65.
0.75	0.02	173.	130.	111.	98.	89.	82.	75.	70.	55.
0.75	0.05	148.	110.	92.	81.	73.	66.	61.	56.	43.
0.75	0.10	129.	94.	78.	68.	60.	54.	49.	45.	34.
0.75	0.20	109.	77.	63.	54.	48.	43.	38.	35.	26.
0.80	0.01	129.	99.	86.	77.	70.	65.	60.	56.	46.
0.80	0.02	117.	89.	76.	68.	62.	57.	53.	49.	40.
0.80	0.05	101.	75.	64.	56.	51.	47.	43.	40.	31.
0.80	0.10	88.	65.	54.	47.	42.	39.	35.	32.	25.
0.80	0.20	74.	53.	44.	38.	34.	31.	28.	25.	19.
0.85	0.01	91.	71.	62.	56.	51.	47.	44.	41.	34.
0.85	0.02	83.	64.	55.	49.	45.	42.	39.	36.	30.
0.85	0.05	72.	54.	46.	41.	37.	34.	32.	30.	24.
0.85	0.10	63.	47.	39.	35.	31.	29.	26.	24.	19.
0.85	0.20	53.	39.	32.	28.	25.	23.	21.	19.	15.
0.90	0.01	67.	53.	46.	42.	39.	36.	34.	32.	26.
0.90	0.02	61.	48.	41.	37.	34.	32.	30.	28.	23.
0.90	0.05	53.	40.	35.	31.	29.	26.	25.	23.	19.
0.90	0.10	46.	35.	30.	26.	24.	22.	21.	19.	15.
0.90	0.20	39.	29.	25.	22.	20.	18.	17.	15.	12.
0.95	0.01	50.	40.	35.	32.	30.	28.	26.	25.	21.
0.95	0.02	45.	36.	31.	29.	27.	25.	23.	22.	18.
0.95	0.05	39.	31.	27.	24.	22.	21.	19.	18.	15.
0.95	0.10	35.	27.	23.	21.	19.	18.	16.	15.	13.
0.95	0.20	30.	22.	19.	17.	16.	14.	13.	12.	10.

TABLE A.3. SAMPLE SIZES PER GROUP FOR TWO-TAILED TEST ON PROPORTIONS
P1 = 0.55

P2	ALPHA	0.99	0.95	0.90	0.85	POWER 0.80	0.75	0.70	0.65	0.50
0.60	0.01	4772.	3559.	2986.	2628.	2361.	2144.	1957.	1793.	1376.
0.60	0.02	4306.	3159.	2622.	2288.	2040.	1839.	1667.	1516.	1137.
0.60	0.05	3666.	2617.	2131.	1833.	1612.	1435.	1285.	1153.	829.
0.60	0.10	3158.	2192.	1752.	1483.	1286.	1130.	998.	884.	606.
0.60	0.20	2620.	1751.	1362.	1128.	959.	825.	715.	620.	397.
0.65	0.01	1187.	891.	751.	664.	598.	545.	500.	459.	357.
0.65	0.02	1073.	793.	662.	580.	520.	470.	428.	391.	298.
0.65	0.05	916.	660.	542.	469.	415.	371.	334.	302.	223.
0.65	0.10	792.	556.	448.	383.	335.	296.	264.	236.	167.
0.65	0.20	660.	448.	353.	295.	254.	221.	194.	171.	115.
0.70	0.01	521.	394.	334.	296.	268.	245.	225.	208.	164.
0.70	0.02	472.	351.	295.	260.	234.	212.	194.	178.	138.
0.70	0.05	404.	294.	243.	211.	188.	169.	153.	140.	105.
0.70	0.10	350.	249.	202.	174.	153.	137.	123.	111.	81.
0.70	0.20	293.	202.	161.	136.	118.	104.	92.	82.	58.
0.75	0.01	287.	219.	186.	166.	151.	138.	128.	118.	94.
0.75	0.02	260.	196.	165.	146.	132.	121.	111.	102.	80.
0.75	0.05	224.	164.	137.	120.	107.	97.	88.	81.	62.
0.75	0.10	194.	140.	115.	99.	88.	79.	72.	65.	49.
0.75	0.20	163.	114.	92.	79.	69.	61.	55.	49.	36.
0.80	0.01	179.	137.	117.	105.	96.	88.	82.	76.	62.
0.80	0.02	162.	123.	104.	93.	84.	77.	71.	66.	53.
0.80	0.05	139.	104.	87.	77.	69.	63.	58.	53.	41.
0.80	0.10	122.	89.	73.	64.	57.	52.	47.	43.	33.
0.80	0.20	103.	73.	59.	51.	45.	41.	37.	33.	25.
0.85	0.01	119.	92.	80.	72.	66.	61.	57.	53.	43.
0.85	0.02	108.	83.	71.	64.	58.	53.	50.	46.	37.
0.85	0.05	94.	70.	60.	53.	48.	44.	40.	37.	30.
0.85	0.10	82.	60.	51.	44.	40.	36.	33.	31.	24.
0.85	0.20	69.	50.	41.	36.	32.	29.	26.	24.	19.
0.90	0.01	83.	65.	57.	51.	47.	44.	41.	39.	32.
0.90	0.02	76.	59.	51.	46.	42.	39.	36.	34.	28.
0.90	0.05	66.	50.	43.	38.	35.	32.	30.	28.	22.
0.90	0.10	57.	43.	36.	32.	29.	27.	25.	23.	18.
0.90	0.20	49.	36.	30.	26.	24.	22.	20.	18.	15.
0.95	0.01	60.	48.	42.	38.	35.	33.	31.	29.	25.
0.95	0.02	55.	43.	37.	34.	31.	29.	27.	26.	22.
0.95	0.05	47.	37.	32.	29.	26.	24.	23.	21.	18.
0.95	0.10	42.	32.	27.	24.	22.	21.	19.	18.	15.
0.95	0.20	36.	27.	23.	20.	18.	17.	16.	14.	12.

TABLE A.3. SAMPLE SIZES PER GROUP FOR TWO-TAILED TEST ON PROPORTIONS
P1 = 0.60

P2	ALPHA	0.99	0.95	0.90	0.85	POWER 0.80	0.75	0.70	0.65	0.50
0.65	0.01	4580.	3416.	2867.	2524.	2268.	2059.	1880.	1723.	1323.
0.65	0.02	4133.	3033.	2518.	2198.	1960.	1766.	1602.	1457.	1093.
0.65	0.05	3519.	2513.	2047.	1761.	1549.	1379.	1235.	1109.	798.
0.65	0.10	3032.	2106.	1683.	1425.	1237.	1086.	960.	951.	585.
0.65	0.20	2516.	1682.	1309.	1085.	923.	795.	689.	598.	384.
0.70	0.01	1127.	847.	714.	631.	569.	519.	476.	437.	341.
0.70	0.02	1019.	754.	629.	552.	494.	448.	408.	373.	285.
0.70	0.05	870.	628.	515.	446.	395.	354.	319.	288.	213.
0.70	0.10	752.	529.	427.	365.	319.	283.	252.	225.	161.
0.70	0.20	627.	426.	336.	282.	243.	212.	186.	164.	111.
0.75	0.01	489.	370.	314.	279.	252.	231.	212.	196.	155.
0.75	0.02	443.	330.	278.	245.	220.	200.	183.	168.	131.
0.75	0.05	379.	276.	229.	199.	178.	160.	145.	132.	100.
0.75	0.10	329.	234.	191.	164.	145.	129.	116.	105.	77.
0.75	0.20	276.	190.	152.	129.	112.	99.	88.	78.	55.
0.80	0.01	266.	203.	173.	154.	140.	129.	119.	111.	89.
0.80	0.02	241.	182.	154.	136.	123.	113.	104.	96.	75.
0.80	0.05	207.	153.	127.	112.	100.	91.	83.	76.	59.
0.80	0.10	180.	130.	107.	93.	83.	74.	67.	61.	46.
0.80	0.20	152.	107.	86.	74.	65.	58.	52.	47.	34.
0.85	0.01	163.	126.	108.	97.	88.	82.	76.	70.	57.
0.85	0.02	148.	113.	96.	86.	78.	72.	66.	61.	49.
0.85	0.05	128.	95.	80.	71.	64.	58.	54.	49.	39.
0.85	0.10	111.	82.	68.	59.	53.	48.	44.	40.	31.
0.85	0.20	94.	67.	55.	48.	42.	38.	34.	31.	24.
0.90	0.01	107.	83.	72.	65.	60.	55.	52.	48.	40.
0.90	0.02	98.	75.	65.	58.	53.	49.	45.	42.	35.
0.90	0.05	84.	64.	54.	48.	44.	40.	37.	34.	28.
0.90	0.10	74.	55.	46.	41.	37.	34.	31.	29.	23.
0.90	0.20	63.	46.	38.	33.	30.	27.	25.	23.	18.
0.95	0.01	74.	58.	51.	46.	42.	40.	37.	35.	29.
0.95	0.02	67.	52.	45.	41.	38.	35.	33.	31.	26.
0.95	0.05	58.	45.	38.	34.	32.	29.	27.	25.	21.
0.95	0.10	51.	39.	33.	29.	27.	25.	23.	21.	17.
0.95	0.20	43.	32.	27.	24.	22.	20.	18.	17.	14.

TABLE A.3. SAMPLE SIZES PER GROUP FOR TWO-TAILED TEST ON PROPORTIONS
P1 = 0.65

P2	ALPHA	0.99	0.95	0.90	0.85	POWER 0.80	0.75	0.70	0.65	0.50
0.70	0.01	4291.	3202.	2688.	2367.	2127.	1932.	1765.	1617.	1243.
0.70	0.02	3873.	2844.	2361.	2062.	1839.	1658.	1504.	1369.	1028.
0.70	0.05	3299.	2357.	1921.	1653.	1455.	1296.	1161.	1043.	752.
0.70	0.10	2843.	1976.	1580.	1339.	1163.	1022.	904.	801.	552.
0.70	0.20	2360.	1580.	1230.	1020.	868.	749.	649.	564.	364.
0.75	0.01	1043.	784.	662.	585.	528.	482.	442.	407.	317.
0.75	0.02	943.	699.	584.	512.	459.	416.	380.	347.	266.
0.75	0.05	806.	582.	478.	415.	367.	329.	297.	269.	199.
0.75	0.10	697.	491.	397.	339.	297.	264.	235.	211.	151.
0.75	0.20	582.	396.	313.	263.	227.	198.	174.	154.	105.
0.80	0.01	446.	338.	287.	255.	231.	212.	195.	180.	143.
0.80	0.02	404.	302.	254.	225.	202.	184.	169.	155.	121.
0.80	0.05	347.	253.	210.	183.	163.	148.	134.	122.	93.
0.80	0.10	301.	215.	176.	152.	134.	120.	108.	97.	72.
0.80	0.20	252.	175.	140.	119.	104.	92.	82.	73.	52.
0.85	0.01	239.	183.	156.	140.	127.	117.	108.	101.	81.
0.85	0.02	217.	164.	139.	123.	112.	102.	94.	87.	69.
0.85	0.05	187.	138.	116.	102.	91.	83.	76.	70.	54.
0.85	0.10	163.	118.	97.	85.	76.	68.	62.	56.	43.
0.85	0.20	137.	97.	79.	68.	60.	53.	48.	43.	32.
0.90	0.01	144.	111.	96.	86.	79.	73.	68.	63.	52.
0.90	0.02	131.	100.	86.	76.	70.	64.	60.	55.	45.
0.90	0.05	113.	85.	72.	64.	57.	53.	48.	45.	36.
0.90	0.10	99.	73.	61.	54.	48.	44.	40.	37.	29.
0.90	0.20	84.	60.	50.	43.	39.	35.	32.	29.	22.
0.95	0.01	92.	73.	63.	57.	53.	49.	46.	43.	36.
0.95	0.02	84.	65.	56.	51.	47.	43.	40.	38.	31.
0.95	0.05	73.	56.	48.	43.	39.	36.	33.	31.	25.
0.95	0.10	64.	48.	41.	36.	33.	30.	28.	26.	21.
0.95	0.20	55.	40.	34.	30.	27.	24.	22.	21.	16.

TABLE A.3. SAMPLE SIZES PER GROUP FOR TWO-TAILED TEST ON PROPORTIONS
P1 = 0.70

P2	ALPHA	POWER								
		0.99	0.95	0.90	0.85	0.80	0.75	0.70	0.65	0.50
0.75	0.01	3907.	2917.	2450.	2159.	1940.	1763.	1611.	1477.	1137.
0.75	0.02	3527.	2591.	2153.	1881.	1678.	1514.	1374.	1251.	941.
0.75	0.05	3005.	2149.	1753.	1509.	1330.	1185.	1062.	955.	690.
0.75	0.10	2590.	1803.	1443.	1224.	1064.	936.	828.	735.	508.
0.75	0.20	2151.	1443.	1125.	934.	796.	687.	597.	520.	337.
0.80	0.01	935.	704.	595.	527.	476.	434.	399.	367.	287.
0.80	0.02	846.	627.	525.	461.	414.	376.	343.	314.	241.
0.80	0.05	723.	524.	431.	374.	332.	298.	269.	244.	182.
0.80	0.10	626.	442.	358.	307.	269.	239.	214.	192.	139.
0.80	0.20	523.	358.	283.	239.	206.	181.	159.	141.	97.
0.85	0.01	393.	299.	254.	226.	205.	188.	174.	161.	128.
0.85	0.02	356.	267.	225.	199.	180.	164.	151.	139.	109.
0.85	0.05	306.	224.	186.	163.	146.	132.	120.	110.	84.
0.85	0.10	266.	191.	156.	135.	120.	108.	97.	88.	66.
0.85	0.20	223.	156.	125.	107.	94.	83.	74.	67.	48.
0.90	0.01	206.	158.	136.	122.	111.	102.	95.	88.	72.
0.90	0.02	187.	142.	121.	108.	98.	90.	83.	77.	62.
0.90	0.05	161.	120.	101.	89.	80.	73.	67.	62.	49.
0.90	0.10	141.	103.	85.	75.	67.	61.	55.	51.	39.
0.90	0.20	119.	85.	69.	60.	53.	48.	43.	39.	30.
0.95	0.01	121.	94.	81.	74.	68.	63.	58.	55.	45.
0.95	0.02	110.	85.	73.	65.	60.	55.	52.	48.	39.
0.95	0.05	95.	72.	61.	55.	50.	46.	42.	39.	32.
0.95	0.10	83.	62.	52.	46.	42.	38.	35.	33.	26.
0.95	0.20	71.	52.	43.	38.	34.	31.	28.	26.	20.

TABLE A.3. SAMPLE SIZES PER GROUP FOR TWO-TAILED TEST ON PROPORTIONS
P1 = 0.75

P2	ALPHA	POWER								
		0.99	0.95	0.90	0.85	0.80	0.75	0.70	0.65	0.50
0.80	0.01	3426.	2561.	2152.	1897.	1707.	1552.	1419.	1301.	1004.
0.80	0.02	3094.	2276.	1893.	1655.	1478.	1334.	1212.	1104.	833.
0.80	0.05	2637.	1889.	1543.	1330.	1172.	1046.	939.	845.	613.
0.80	0.10	2275.	1586.	1272.	1080.	940.	828.	734.	652.	454.
0.80	0.20	1891.	1271.	993.	826.	706.	610.	532.	464.	304.
0.85	0.01	803.	606.	513.	455.	411.	376.	346.	319.	251.
0.85	0.02	727.	541.	453.	399.	359.	326.	298.	273.	211.
0.85	0.05	622.	452.	373.	325.	289.	260.	235.	214.	160.
0.85	0.10	539.	383.	311.	267.	235.	209.	188.	169.	123.
0.85	0.20	451.	310.	247.	209.	181.	159.	141.	125.	88.
0.90	0.01	329.	251.	214.	191.	174.	160.	148.	137.	110.
0.90	0.02	298.	225.	190.	169.	153.	140.	129.	119.	94.
0.90	0.05	257.	190.	158.	139.	125.	113.	104.	95.	74.
0.90	0.10	224.	162.	133.	116.	103.	93.	84.	77.	58.
0.90	0.20	189.	133.	108.	92.	81.	73.	65.	59.	44.
0.95	0.01	167.	129.	111.	100.	92.	85.	79.	74.	61.
0.95	0.02	152.	116.	100.	89.	81.	75.	70.	65.	53.
0.95	0.05	131.	99.	84.	74.	67.	62.	57.	53.	42.
0.95	0.10	115.	85.	71.	63.	57.	52.	47.	44.	34.
0.95	0.20	98.	71.	59.	51.	46.	41.	38.	34.	27.

TABLE A.3. SAMPLE SIZES PER GROUP FOR TWO-TAILED TEST ON PROPORTIONS
P1 = 0.80

P2	ALPHA	0.99	0.95	0.90	0.85	POWER 0.80	0.75	0.70	0.65	0.50
0.85	0.01	2849.	2133.	1795.	1584.	1426.	1298.	1188.	1091.	844.
0.85	0.02	2574.	1897.	1580.	1383.	1236.	1118.	1016.	927.	703.
0.85	0.05	2196.	1577.	1290.	1114.	984.	879.	790.	713.	521.
0.85	0.10	1896.	1326.	1066.	907.	791.	698.	620.	553.	388.
0.85	0.20	1578.	1065.	835.	697.	597.	518.	453.	397.	264.
0.90	0.01	646.	490.	416.	370.	335.	307.	283.	261.	207.
0.90	0.02	586.	438.	369.	325.	293.	267.	245.	225.	176.
0.90	0.05	503.	367.	305.	266.	237.	214.	195.	178.	135.
0.90	0.10	437.	312.	255.	220.	195.	174.	157.	142.	105.
0.90	0.20	367.	254.	204.	174.	151.	134.	119.	107.	77.
0.95	0.01	254.	195.	168.	150.	137.	127.	118.	110.	89.
0.95	0.02	231.	176.	150.	134.	121.	112.	103.	96.	77.
0.95	0.05	199.	149.	125.	111.	100.	91.	84.	77.	61.
0.95	0.10	174.	128.	106.	93.	84.	76.	69.	64.	49.
0.95	0.20	148.	106.	87.	75.	67.	60.	55.	50.	38.

TABLE A.3. SAMPLE SIZES PER GROUP FOR TWO-TAILED TEST ON PROPORTIONS
P1 = 0.85

P2	ALPHA	0.99	0.95	0.90	0.85	POWER 0.80	0.75	0.70	0.65	0.50
0.90	0.01	2176.	1634.	1378.	1219.	1099.	1002.	918.	845.	658.
0.90	0.02	1968.	1456.	1215.	1066.	955.	865.	788.	721.	551.
0.90	0.05	1682.	1213.	996.	862.	764.	684.	617.	558.	412.
0.90	0.10	1454.	1023.	826.	705.	617.	547.	488.	437.	312.
0.90	0.20	1213.	825.	651.	546.	470.	410.	360.	318.	216.
0.95	0.01	466.	356.	304.	272.	247.	227.	210.	195.	157.
0.95	0.02	423.	319.	271.	240.	217.	199.	183.	169.	134.
0.95	0.05	365.	269.	225.	198.	178.	162.	148.	136.	105.
0.95	0.10	318.	230.	190.	166.	148.	133.	121.	110.	84.
0.95	0.20	269.	190.	154.	132.	117.	104.	94.	85.	63.

TABLE A.3. SAMPLE SIZES PER GROUP FOR TWO-TAILED TEST ON PROPORTIONS
P1 = 0.90

P2	ALPHA	0.99	0.95	0.90	0.85	POWER 0.80	0.75	0.70	0.65	0.50
0.95	0.01	1407.	1064.	902.	800.	725.	663.	610.	563.	445.
0.95	0.02	1275.	950.	798.	704.	633.	576.	527.	484.	376.
0.95	0.05	1093.	796.	659.	574.	511.	461.	418.	381.	288.
0.95	0.10	949.	676.	551.	474.	418.	374.	336.	303.	223.
0.95	0.20	796.	550.	439.	373.	324.	286.	254.	227.	161.

08939	53632	41345	65379	20165	32576	13967	90616	17995	92422
92578	23668	08801	39792	59541	99117	58830	60923	36068	68101
83994	91054	90377	22776	23263	34593	98191	77811	83144	98563
43080	71414	40760	01831	44145	48387	93018	22618	98547	87716
39372	46789	26381	37186	85684	79426	05395	17538	56671	82181
83046	58644	04452	98912	53406	30224	00687	32099	86414	29590
99808	32539	96961	88917	60847	64826	41332	64557	15354	11111
28478	70870	68912	75644	33648	21097	23745	52593	01849	37760
09916	19651	28659	95093	12626	19919	05879	56003	83100	94572
19537	66067	20569	28808	87722	67059	12851	73573	25776	92500
23013	05574	26320	07754	09642	88068	41626	57139	68199	94938
55838	80585	80967	60540	34528	62310	63106	17843	39104	74036
92279	87344	93556	75233	09394	79265	91047	32891	77925	71530
27850	23332	89336	26026	52130	78544	02090	05645	15060	39550
01760	54605	11794	79312	69728	04554	99775	57659	47981	68954
81889	70751	87501	88247	41966	57574	67745	88304	20118	25964
74722	14654	15425	60665	25162	04987	03467	75915	24282	62456
56196	75068	44643	92240	51651	79743	13598	63901	61020	91003
96842	62021	00543	45073	65545	87612	35765	26079	34589	72821
25619	98328	59393	71401	93871	20611	78830	87477	15390	05044
91746	05084	04781	82933	54564	80986	94843	40178	87483	63288
92384	84706	76778	98313	98875	08427	60687	88272	83448	06237
86390	62208	95735	14535	25591	22730	06059	31786	36181	31016
60458	83606	57510	92609	38061	94881	26736	06489	98303	31419
03783	39922	05489	73630	92379	91602	18193	84741	44704	05558
31011	36035	37113	98362	56149	51634	04468	62096	32361	35301
20555	05621	48728	41776	12101	96615	70781	55151	93876	66892
56466	36766	12400	43510	49456	05140	85736	68155	37306	10438
26875	67304	61950	65962	38223	35676	70043	99178	64677	95457
90648	84770	92791	93814	27760	22232	83545	01183	55188	20482
26197	72840	01264	52019	00739	36259	10905	39097	36437	66743
72522	34445	53975	13840	97262	59007	78685	41044	38103	59216
12370	41270	36290	46307	51230	90614	82613	80148	37371	02895
81028	60112	31415	47478	02131	85480	93699	92876	13958	47867
61573	38634	77650	18189	10283	97999	95442	90657	84963	93863
98511	46300	91199	30492	62159	98525	31710	03540	35844	83200
76606	10834	75548	55779	54744	26450	66001	57949	53685	00567
20237	16311	15733	47599	43998	35594	17577	85113	52487	48900
21022	86025	26951	87480	82317	06580	98627	32536	07573	52612
47512	11564	41777	46581	03492	01722	78900	57901	37307	02727
80598	59041	28861	41793	91007	69907	00376	73086	35132	53014
01892	34226	88327	21926	36607	22307	04376	25491	13563	51955
89657	70349	15176	57916	10911	44218	67108	04678	24097	02476
97983	65616	11841	80504	76452	34176	16986	94328	13091	29592
59727	92033	14654	59622	25844	18460	78162	02832	13528	55683
12340	72894	26303	01771	73895	27432	99536	50328	06141	83886
48049	33318	67463	04914	22316	89663	37132	15825	60759	22131
85953	16537	25639	05004	99269	50577	10036	05022	39800	93605
03426	78111	37828	23967	03350	04397	96227	37787	60680	23993
97837	71085	45973	36073	02680	91425	24425	23725	22521	21601

195

26916	67086	60270	57846	04646	07258	01734	45079	54869	23505
47205	71678	05222	86233	70398	46287	44139	48247	92230	19157
84869	36794	56943	10512	50582	08884	98068	08447	68071	32397
81740	98868	57546	55461	14850	89946	06024	26626	05543	93616
11808	28306	63559	26600	87569	86007	27922	93468	09509	15841
09464	14219	00130	72813	35704	58905	32091	62397	85560	51783
40656	77886	01411	07490	32240	26028	66002	61762	76551	03442
31693	59176	69817	86317	89547	60424	56618	95888	65770	31622
97799	02197	32987	78146	71992	28633	23868	85504	98216	19756
34590	29732	67082	34899	05654	19830	68088	30054	67535	34721
00504	90537	38681	17248	55362	76935	63352	87699	56022	46835
76814	39363	44851	14836	85357	78617	03482	13336	48678	72047
94171	16606	52092	63096	09752	90644	56092	20751	19678	31311
10758	82747	99662	53243	22501	55820	32406	92052	60659	35477
66933	82305	91425	07804	24003	73777	26634	95806	35126	48503
74883	12771	02671	01090	82498	85176	68569	44827	51844	07616
79102	06066	24478	92267	33300	69392	16652	75381	02415	36065
94649	43308	08005	58253	77473	40559	46096	11540	54375	22388
44952	68217	04728	14414	43931	33854	07744	41771	80933	09655
29531	90289	75949	43091	75005	62207	98196	29316	92128	88918
04355	25867	16008	63243	35388	43138	40330	53741	59469	73144
03640	63541	48488	19060	77959	96217	75666	88042	47261	91184
21749	09836	63276	91133	77308	43654	66146	03991	28629	35848
57425	21919	14688	90852	12918	59833	42736	17916	22868	75963
49962	16108	46986	36939	98761	60113	71822	89915	93090	49299
57985	04214	03417	82576	64699	45011	87770	21525	39212	41547
96374	04318	58540	12375	47382	72917	11063	10129	61201	76044
05457	48338	40916	10453	94473	72759	86299	62959	01064	39749
28918	15769	34348	64162	75841	77582	82921	99286	49425	02973
73010	11300	10710	62560	78969	10771	53899	26454	73627	03681
59435	23480	05967	24479	93169	38697	93658	13676	39128	11680
17929	02455	53366	98097	08284	66830	26423	57062	04563	13822
41862	26768	83848	62175	18414	50906	39708	80097	23206	18358
01294	42540	43590	78681	79771	70501	05062	95860	29602	14866
22775	02858	12165	47273	74148	19427	49227	20518	80065	95722
53747	60983	82171	44180	25536	55599	03762	22186	99253	95841
81766	28025	32247	41257	57319	72602	19740	65016	12435	89463
24862	44004	40269	45574	11018	55941	36479	15404	64110	46027
99169	80770	92093	25630	24942	18977	89382	65496	88534	41734
53600	45992	93546	47348	42169	26882	81774	48703	56244	99137
88627	39523	39496	81268	32137	61411	79234	22696	23073	34171
97367	76657	83638	11912	18723	05129	62265	27431	04195	78294
32005	87382	36246	31037	60009	80722	44244	38968	35608	62938
57154	76478	44478	78561	71064	19331	76406	84452	19058	54278
54146	36375	30932	58210	70875	01355	70257	09341	23730	58309
36283	92917	30953	49460	18185	63965	20121	45041	89156	29563
74973	83767	27843	13152	28328	51597	54624	63371	88603	61277
70237	69924	87413	95159	84237	48986	35781	73808	20817	60630
76426	12882	89455	20792	19655	30803	07915	70264	50346	69701
65088	63220	93521	92145	11180	37773	26018	16150	62735	31062

```
21755  50969  10016  01373  18088  96168  14217  19786  90759  66476
82024  93860  24943  04919  05019  85844  69890  46740  51431  87922
64649  96595  97725  16988  22404  81529  87537  91453  60886  42239
05455  52581  66391  25111  53143  92863  78886  37547  15306  53911
85711  29066  02999  56394  11372  60689  61784  24499  90934  25106
32230  67428  14496  80119  50249  80419  30275  57878  74784  27806
93773  12383  30343  70604  50537  67783  51863  01132  40022  29939
19436  47161  08039  23786  70362  08094  15302  18963  76059  85683
29564  06230  71308  71770  88850  87166  23344  55564  23287  39647
28294  12945  23018  21604  22457  40306  39721  75568  95922  95419
09211  96490  96042  07837  82647  25343  08236  21325  53823  31010
01652  30822  70058  42947  27160  76437  14177  97132  55193  56972
45091  57793  40937  25483  84462  77419  04356  29363  36969  57549
12567  57462  31667  72844  52056  56741  71936  20944  78241  80949
81524  60599  29872  33841  34193  00587  95783  69415  54442  01910
21482  11696  76840  55775  43085  56535  51444  99849  36099  17950
82810  35306  66543  81499  90106  07145  31914  27172  75808  10295
79498  84331  90497  84000  89528  81166  81247  56983  10673  51195
11109  05896  35392  59285  37186  89548  02607  09712  34804  21413
15244  98745  55271  42923  60096  74268  04743  60039  17547  64932
31666  05605  48629  41332  10329  89982  46927  71723  07996  02466
42826  34764  23143  25983  47607  51791  82282  27570  24876  01128
82881  87130  76850  76921  69879  26981  32973  55008  33291  04669
28391  28322  14413  31579  59754  74317  08112  79815  05879  16938
48719  39869  00739  45610  67010  91567  46312  53765  05780  50798
52763  59397  81517  54521  93475  70156  79661  46562  62420  55458
70967  26680  21377  88141  36450  25424  24495  18149  88435  67268
07692  40737  75193  84524  30406  21722  56673  44542  57189  42256
38832  52688  66638  25632  54050  93604  75178  08625  75145  73248
63182  19854  50484  24217  90941  27692  47680  36849  91973  20190
93388  78611  31175  79544  96694  64262  15325  13587  43599  39302
43423  06816  50091  39199  84373  53446  61320  86900  69517  35003
84358  91122  87506  04936  42059  07924  69016  42775  35505  28060
48808  40305  02561  52614  92636  82287  60001  19417  76491  84195
16750  42742  05696  49496  45709  28786  61339  08953  01668  29427
61552  04467  72828  96765  25138  79942  53404  00946  25034  41690
24829  73764  19122  50857  33043  40546  45884  10391  49390  02819
72401  09034  02594  34257  82193  84846  69338  52408  90406  70765
10932  28706  73841  84692  43581  99260  03325  26610  29737  38927
65930  45238  78052  61167  64536  36708  39425  06176  82227  37781
86639  79801  99050  76091  65094  05740  48597  39918  02130  53520
01947  29996  62454  04755  66442  55854  37146  20187  86811  39179
53770  70012  36138  86720  95077  89978  84171  95222  13796  25774
30475  50884  31026  28195  89935  85855  05715  61588  18092  54261
62739  88081  63832  90260  67072  90095  36914  10629  31549  93630
77844  24386  45720  54845  17591  11938  13307  72402  14648  36263
85747  88110  56936  48625  79327  38333  16052  51315  63422  18693
41361  79928  07316  24546  81431  67669  73127  29744  20315  54192
42317  66171  16169  01470  63300  66571  29722  79191  07644  33148
30088  82194  22366  94453  65418  42430  65820  13046  85896  97958
```

14679	06451	06588	81467	29514	43874	97618	90837	66459	81223
89204	04220	22479	96891	79032	02169	01727	90834	29465	59280
90491	09422	42113	67877	28245	58572	97229	56225	62283	67668
45465	63773	73453	66756	39456	35932	60485	39521	92761	00876
70407	14550	53641	90888	62562	57373	81180	70722	93714	25780
11254	35507	11749	50931	99843	40677	45472	06738	31180	26435
77992	30030	78254	99249	91207	54003	03149	32871	68881	17444
84733	51402	45811	12247	61182	56872	90592	31698	74858	63867
09476	82052	07023	03671	58821	19882	89900	60456	53637	77719
83577	01987	59725	40464	05258	67372	56907	35085	98351	74336
60855	96100	28905	08526	91010	69331	96888	57348	72097	16448
49813	50842	56729	82794	86204	72075	56610	90983	36404	99578
69832	22786	08230	44306	91759	51802	84976	43633	97016	89399
63245	74878	80059	06446	18147	50861	41846	93322	83316	59991
10105	20707	33291	13385	80687	63653	55732	61518	67517	51439
00977	42906	48644	05707	96448	27318	95873	29376	13401	16019
75501	08838	73515	61548	07645	91940	82831	69523	71658	04241
80525	44978	45143	66055	90038	45735	64065	72771	60040	05302
91446	00959	82739	87803	82164	02753	77038	37448	31342	51018
24031	85017	93826	97797	42345	73895	62266	08538	90827	68122
91285	34606	86074	04984	55238	86574	22300	54630	27078	70794
81062	49226	65798	51749	18313	44134	99983	02693	16310	73623
94945	07061	87861	63618	90956	73170	20417	63972	00075	24697
47508	62774	49067	29437	35021	45184	55918	28848	69823	59308
27440	30866	38231	51593	65475	28517	81818	34260	69189	06796
18072	47268	20958	00335	13388	77308	43355	64352	95916	96329
14722	21374	95744	82090	30844	46254	99924	83255	00209	51956
09853	91508	60367	38631	88481	83202	02881	68468	84877	93045
94373	28832	23632	82309	81160	46174	46608	64077	64988	13237
94526	48016	37362	92027	15906	67185	59957	55076	90844	49373
78640	27483	57136	95467	58574	92235	26245	27355	27918	21312
76610	67852	17624	95070	11800	15172	84826	72409	71016	74412
07321	74221	79429	08584	36645	42613	25865	71671	97244	38423
55337	86213	19237	39510	63919	27925	92274	02322	83458	79852
27987	49253	43638	18546	18533	44280	98884	94782	78736	19373
07610	71301	59319	14196	51309	80085	18726	91587	80992	61663
41047	91314	78463	48950	87529	84618	19947	58119	69184	92035
29552	48994	27991	27000	10082	17489	14188	27732	38692	82566
47166	39897	14884	30234	47445	12559	48351	77069	27255	69909
74159	15772	27203	21269	82781	05262	86541	71890	52467	67787
34518	97021	71462	55582	90333	41760	37564	49543	59181	09194
22533	52449	11893	99313	88836	39198	35665	61203	46236	26586
43385	21033	35736	25113	29055	48307	28348	35323	56803	22908
26224	51169	70990	65422	53620	32916	14914	93239	25205	12074
45601	64938	79216	90857	32195	75453	62960	98682	25443	64524
58151	68188	85768	00919	33595	93299	57440	04948	12724	31809
76340	71754	43465	15001	98816	57885	57961	26798	39138	93645
09625	14943	03037	83731	75049	96716	04849	58650	08254	21673
55746	39415	34777	53925	10558	78021	73106	36443	60700	36211
70966	99897	60680	65010	43936	78523	75711	47553	03918	95528

198

37906	67684	64945	80515	98580	66844	13841	81451	64133	51736
33214	33660	03036	15271	64298	48348	11405	33299	97141	83158
24673	99616	75635	57261	62852	61758	04883	73473	96889	20071
48424	09907	23621	52564	02788	43652	36819	28046	36901	68989
81824	70042	83835	72632	81280	33986	72398	28508	19464	60208
86072	74558	72701	65183	36787	34078	73381	33580	41052	44089
95065	73583	85914	53236	46186	97991	72273	51718	59845	93607
23036	95754	67198	41400	43613	89077	41942	49956	22261	83956
03385	64705	57768	64255	65622	15430	01983	73023	20295	64558
04689	47109	40454	18734	48321	21315	93002	66173	60023	64576
47246	02296	88556	10674	67034	06143	33545	45982	73986	30075
14570	16742	69321	65249	67601	18364	01776	45378	56279	29273
69122	51274	85540	51772	40770	78669	05078	22446	88971	31817
90157	54589	16114	05382	87269	75173	65610	17102	12127	18837
03879	53737	87508	41417	60925	92795	08442	15497	17004	62546
22236	70503	22890	02810	10852	39816	41230	89031	63119	77428
96497	82130	24298	06620	21037	66636	10484	63177	84701	39611
75355	95630	95584	12836	16760	85032	59349	90801	10663	46768
84645	86953	59912	76888	22124	40748	45370	05479	24547	97967
66878	19564	15482	16818	61566	18034	54106	62327	87010	61112
83586	51499	56724	76848	50567	11768	15509	87138	83245	15223
42133	15790	15539	51120	66864	41105	44854	99171	91340	55499
10936	66119	98287	94653	83328	48089	38583	98693	44910	81224
83154	67905	59041	43101	27232	75480	07791	67427	34439	99786
88765	34520	08226	38677	58026	00065	78152	68330	06605	24659
88509	09116	58112	66633	76782	60996	74937	00650	29469	70960
60539	24597	02726	94980	45345	17247	95374	17014	43723	09207
61732	87526	69563	29647	51811	44040	97946	91308	66335	76237
60406	76296	14122	98066	61291	85153	59295	89395	02711	11708
45845	69699	05584	06215	87034	66264	14275	89270	07145	39440
72327	79003	23073	27408	56790	94062	53259	72995	58638	94869
41471	95004	96783	25663	82927	66588	53186	19823	40260	63150
16553	30969	36833	42272	22134	52349	14887	18182	75107	87002
46044	93244	45068	31210	81646	08985	19099	33722	30442	79154
00942	93267	51118	67305	43765	56843	47168	71421	04014	81289
51611	78061	03866	20642	89054	48544	89778	01766	02593	99660
74620	50774	33064	41415	50910	32721	38133	34307	62641	67084
38731	28630	23199	81522	80336	48317	66878	66416	96592	81803
21487	92695	62781	42431	89552	55433	26629	60878	25607	05738
03962	72124	97090	33420	26704	59442	16319	62932	30719	17924
31070	25105	70997	00037	61248	67153	51682	05715	69150	63466
58445	79472	50823	89685	80704	77058	36012	22548	11179	64137
84211	28033	10300	09500	64296	00270	22957	35313	05259	13737
35087	86887	05537	51242	57611	84488	88429	70179	25215	19675
91108	69575	97474	58666	74734	20403	49813	15245	43150	21696
41824	55683	57679	44922	50418	98213	35513	29157	55322	69516
19196	89530	64420	80747	04696	01452	66444	85596	15581	23122
98495	82872	10780	18824	15923	26125	13436	45495	52042	02791
48367	65082	55187	45381	75601	45172	90624	37190	07520	10413
94793	75039	97093	07207	69403	51555	84698	44192	02867	19475

91046	70999	06579	00484	43692	57795	53540	01085	24645	38105
06824	97998	26567	77415	25388	55594	05067	30055	34726	37855
14595	46870	49867	77370	15416	96161	38219	63861	39195	60419
09751	14737	00661	71330	22026	90185	42872	45571	87570	15285
03574	83877	71094	71667	90150	95901	64049	21185	50667	13340
24033	24135	28513	53860	66540	14499	88130	98287	96551	94717
99341	43590	67468	12497	67768	94135	54895	82151	98848	53733
32760	12964	82939	80956	39285	07097	89021	70247	20290	89516
54491	21296	37353	32456	58553	59215	28308	36913	66700	67981
07583	33673	33787	99662	93888	66367	53206	21934	52744	19057
39647	66364	41356	50860	32955	39987	43329	00091	10580	62660
80732	20453	96128	92692	90996	11743	51493	03267	56168	07600
40090	72135	71999	82780	48686	47093	44383	42461	55315	49745
00632	56081	30796	80045	03103	98211	61335	84114	52662	58949
79731	47847	69497	86357	92669	78799	38769	23422	91608	38848
08614	02049	34765	90145	27982	56589	87690	16839	11823	19391
09935	85090	21118	60897	75323	03859	45248	36755	13296	48981
74216	04471	58876	13015	48205	72092	98703	43385	71982	41419
00677	31291	81644	08248	14690	13905	51216	82150	31951	52357
26580	88272	90406	41986	74259	13682	13754	59388	32537	60731
71552	82733	52426	69959	47917	57872	15979	75022	06320	62721
19442	52160	37978	58426	08753	26686	81330	47810	54884	99014
00122	09604	56525	52716	07567	70954	57616	07110	24118	80713
67216	76877	56318	46014	69224	01212	84256	94624	09438	05009
45110	25576	47459	54570	00287	10586	60938	70350	73650	08754
89671	59232	48355	57037	07029	28837	09759	99012	06245	46358
21941	14417	89031	04431	25308	11970	44044	56533	42796	47979
02705	84418	82163	33219	59841	60076	21881	90604	46695	64728
68115	26133	43759	27355	70302	75614	20963	45249	82820	89682
92707	49102	60251	19581	75228	75131	73738	66245	33821	06721
35939	55142	07399	48110	22069	99420	97897	92602	74539	13811
12019	47810	78689	41839	42832	80434	97117	58792	78698	43063
50092	12981	27059	45518	29575	67789	40553	33217	34323	06982
32986	35078	13588	65822	72642	43450	06917	50448	40435	88575
67536	08040	40407	70084	56838	10269	50074	08019	97450	12526
98106	75894	72414	51431	56860	78280	57941	43121	37253	35425
77276	44827	73475	37404	63144	42226	85056	30301	16297	25073
03764	96921	47652	13621	52855	94537	91525	98316	66168	12161
77455	55276	34556	09855	48125	00049	67169	02566	10876	42160
55073	50995	10314	02925	24721	22000	09511	59060	68761	81026
67299	74565	41692	79065	99163	83388	07859	96658	09217	85375
29298	07412	80784	17996	80921	23560	13066	66357	80544	86050
91407	73988	21260	61665	78651	16919	93657	09667	15090	03532
85379	80488	14514	62693	45528	08934	43846	82672	01415	64436
73877	63341	15148	20821	88589	44145	67572	08120	40576	70372
57044	08909	40062	60187	00559	61672	65001	34955	24717	33706
79784	75348	34030	26050	50030	65726	44081	72952	40980	89306
67019	98358	86971	36607	36901	91939	19522	89680	62379	67151
41496	44616	94227	63819	34868	34838	95217	57756	89580	17676
99834	39917	40995	86715	51336	27577	03438	72434	03659	70046

```
87340  93621  75668  11412  87464  22069  45241  72818  29736  23057
70713  16765  64165  34106  27152  55954  91352  44525  44980  69154
10102  38228  38442  86601  73625  62341  11417  07431  41828  84087
28072  11642  17204  98450  35856  29088  51820  49122  28351  28007
12884  25234  35447  85574  94423  96364  28378  02991  62542  48328
27094  27605  21783  82254  97475  44559  90078  39433  25896  00475
69789  14454  58626  21664  02351  19124  93582  89374  94004  59659
11915  34559  00117  89667  36950  14694  55609  01408  07963  35106
38964  17831  56305  77347  57337  47900  71366  97092  40254  67697
43892  54073  29411  89805  74131  36540  52057  83475  32340  42757
65481  08075  59117  82023  60088  22314  93095  57736  08564  31758
13466  94974  48647  37115  84859  75118  86979  45810  92049  40000
11558  09341  52027  28088  00284  48909  90897  05195  13099  31837
73126  52226  55216  61265  70639  72451  98949  41632  59251  80815
51628  82433  29946  37772  57118  02761  02502  90157  18424  99129
28958  81585  28890  39072  74422  94883  99498  43036  62730  89054
69755  17039  74438  93269  89674  98623  84667  20393  60353  78578
28290  62543  20640  60954  79960  31174  67404  23853  36480  04202
96886  43500  89025  42647  54653  44096  72699  39322  81643  35955
80940  62044  43802  04416  32270  53875  32820  12044  76883  52900
25451  76608  30584  94033  88940  87339  23568  55358  20032  21972
51540  11490  05081  27069  16685  56486  88750  24122  45978  58768
38809  03936  74336  10590  94518  71798  80119  34531  86118  05922
60470  09522  12897  91680  34003  78900  67368  94108  58328  02998
93030  31200  49927  18762  63223  10480  93871  68905  68583  91355
30880  63084  00584  35748  09225  33618  18680  09517  88981  48227
88533  97152  86114  42311  78843  92251  43919  33256  04265  26280
19299  79270  01925  98119  71388  45254  29028  66878  40017  38202
69054  70503  01526  74633  34061  32661  89417  42554  50565  20407
67352  20450  16530  15129  41996  15818  16940  59277  03202  85715
85468  41376  79039  01850  99744  81812  93169  22705  97709  81907
12057  35180  02566  98772  69539  28280  43827  08444  56220  61323
61951  19798  61229  89189  84075  01746  53801  07092  58342  86217
92224  77389  34320  09414  47608  00915  77016  53859  30015  95353
01982  53711  04423  43139  19021  25871  84038  71386  71973  89367
88443  26351  62114  35523  54106  04932  42631  11399  84712  05686
71700  79026  28855  61893  11658  12915  72563  19139  61766  98345
34173  19937  12062  92936  49051  57880  05821  14005  31644  63815
98089  14200  02395  86565  97833  07913  66981  30666  81164  10981
35414  13649  63167  56163  68467  05339  15824  46897  38963  11698
19522  11849  95389  65695  35664  22725  15372  87703  87868  37883
36480  77931  39265  34212  51883  03389  53384  89802  58350  41882
26138  79888  44088  45532  76396  48583  03931  86340  82657  18881
69367  46269  53313  03455  40910  14365  18002  78724  10321  53409
27566  84710  60165  98597  50090  13172  28217  50750  50546  46520
24208  26566  41518  10015  86420  28387  92538  99744  65618  96011
85497  48885  23833  03027  03664  94744  35480  60186  41792  09076
78331  88300  24816  54195  01822  23175  22655  27351  60205  15074
48594  55901  98052  85207  28770  05754  75593  01764  30247  65604
21400  37964  35184  69430  99920  74647  48604  19798  81347  09895
```

201

27247	74427	01337	38176	17021	58543	98070	61531	86549	65516
14156	95285	44308	08275	50878	30794	26858	84002	62289	17713
45673	14618	76646	28312	80058	25534	32686	66307	03660	25200
18256	82731	32083	47921	98771	61337	79083	22463	23032	36026
08862	28939	93878	02811	71963	06475	91182	88814	12247	74153
34694	40786	32471	27744	74229	95669	05957	74718	94691	95680
21859	70031	23453	10432	51516	15202	27567	28584	23404	83160
88328	81523	94184	31389	40680	61577	03337	65833	64959	97256
98901	18100	18490	48034	21791	98440	94518	81149	36224	87001
95993	92944	93729	25874	11685	37237	18258	74414	82159	23219
99886	90343	43080	45388	84604	99135	33371	08008	47709	14181
55707	06608	38286	70240	76861	29006	82287	32662	55387	38363
31696	44902	84149	00774	47296	76811	35197	19884	02527	36204
94480	41042	95933	06218	73908	87481	59713	70946	88258	91030
51854	91851	84421	79859	19367	97465	10491	85755	20110	48867
12205	33427	90714	43441	44220	74345	48089	19423	83736	27611
12035	23714	33964	90354	36447	05492	04934	80168	36601	98093
59152	72070	00051	51676	09597	92497	68607	79172	57563	32830
78914	78012	57844	44952	49118	90138	98763	81334	99134	62792
84548	42156	91999	72590	07545	91955	83829	75373	97778	08310
69856	44345	37359	25052	14076	58986	27231	32509	49973	07253
93760	97283	39850	63554	22674	64053	80248	05012	07838	01917
40959	28502	02379	57758	25133	30976	59655	79147	37982	15567
51562	69272	51570	85970	51690	36403	53207	91615	70820	00385
64929	86104	32268	18664	21571	61451	74567	94342	94943	20582
69002	28768	51591	50633	39475	81155	31652	59516	72230	97730
36308	38281	02907	72917	11332	11740	68452	05049	14222	39893
11359	09112	52439	32620	23768	49023	80229	40159	18894	51933
41544	81869	17314	67064	46558	75767	35582	31585	69270	31356
64702	06004	53711	68222	25934	41607	16231	22923	91454	42414
31398	06654	57345	84180	88969	76195	56445	52913	09476	80635
98527	65438	05890	46391	25333	34481	78886	62987	67946	40786
33203	32139	94004	74772	02591	42598	32264	10207	70859	33289
62005	72426	76511	07230	54774	63575	88484	58723	55982	07384
40465	76337	93835	75973	11321	84168	03118	61194	39104	83876
51313	52998	56165	60010	54571	87333	32852	11115	71019	26079
17303	69107	58913	31510	58840	69451	87141	97787	02447	34599
85562	01985	41848	33221	22690	37149	18677	77720	98226	89880
55239	22513	37930	24957	08372	25614	78337	39488	31896	35980
28818	49087	35155	69148	98487	68591	25162	33651	75444	49804
19825	70717	45870	38769	19778	69745	40464	15079	26297	22066
95718	75713	92817	75483	17542	25903	97538	52096	34734	39538
24624	91901	29784	51594	41504	84679	34535	45096	59754	52661
78177	95113	67082	46472	75090	32286	17907	16865	40027	88378
70022	24726	18150	86370	54867	51867	17396	37574	68877	75099
30696	08284	73441	66084	35537	18463	90940	79472	58367	34949
84393	91813	91338	21704	08183	53759	48903	09586	17386	18038
51751	48159	23192	05719	25583	02027	81915	73246	02237	54207
05106	42768	10657	79027	78247	58237	45199	47058	75553	29795
98790	24578	58358	28940	48422	30065	44593	96972	80495	10221

202

LN(10)=2.30259.
IF X IS GREATER THAN 10, EXPRESS X AS THE PRODUCT OF A NUMBER, Y, BETWEEN 1
AND 10 TIMES 10 RAISED TO A CERTAIN POWER. IF N IS THE POWER OF 10, THEN LN(X)
= LN(Y) + N*LN(10). FOR EXAMPLE, 974 = 9.74*(10**2), SO THAT Y = 9.74 AND N= 2.
THEREFORE, LN(974) = LN(9.74) + 2*LN(10) = 2.27624 + 2*2.30259 = 6.88142.
IF X IS LESS THAN 1, EXPRESS X AS THE RATIO OF A NUMBER, Y, BETWEEN 1 AND
10 DIVIDED BY 10 RAISED TO A CERTAIN POWER. IF N IS THE POWER OF 10, THEN LN(X)
= LN(Y) − N*LN(10). FOR EXAMPLE, 0.974 = 9.74/(10**1), SO THAT Y= 9.74 AND N=1.
THEREFORE, LN(0.974) = LN(9.74) − LN(10) = 2.27624 − 2.30259 = −0.02635.

X	0.00	0.01	0.02	0.03	0.04	0.05	0.06	0.07	0.08	0.09
1.0	0.00000	0.00995	0.01980	0.02956	0.03922	0.04879	0.05827	0.06766	0.07696	0.08618
1.1	0.09531	0.10436	0.11333	0.12222	0.13103	0.13976	0.14842	0.15700	0.16551	0.17395
1.2	0.18232	0.19062	0.19885	0.20701	0.21511	0.22314	0.23111	0.23902	0.24686	0.25464
1.3	0.26236	0.27003	0.27763	0.28518	0.29267	0.30010	0.30748	0.31481	0.32208	0.32930
1.4	0.33647	0.34359	0.35066	0.35767	0.36464	0.37156	0.37844	0.38526	0.39204	0.39878
1.5	0.40546	0.41211	0.41871	0.42527	0.43178	0.43825	0.44468	0.45107	0.45742	0.46373
1.6	0.47000	0.47623	0.48243	0.48858	0.49470	0.50077	0.50682	0.51282	0.51879	0.52473
1.7	0.53063	0.53649	0.54232	0.54812	0.55388	0.55961	0.56531	0.57098	0.57661	0.58221
1.8	0.58779	0.59333	0.59884	0.60432	0.60977	0.61518	0.62058	0.62594	0.63127	0.63658
1.9	0.64185	0.64710	0.65232	0.65752	0.66269	0.66783	0.67294	0.67803	0.68310	0.68813
2.0	0.69315	0.69813	0.70310	0.70804	0.71295	0.71784	0.72271	0.72755	0.73237	0.73716
2.1	0.74194	0.74669	0.75142	0.75612	0.76081	0.76547	0.77011	0.77473	0.77932	0.78390
2.2	0.78846	0.79299	0.79751	0.80200	0.80648	0.81093	0.81536	0.81978	0.82417	0.82855
2.3	0.83291	0.83725	0.84157	0.84587	0.85015	0.85441	0.85866	0.86289	0.86710	0.87129
2.4	0.87547	0.87963	0.88377	0.88789	0.89200	0.89609	0.90016	0.90422	0.90826	0.91228
2.5	0.91629	0.92028	0.92426	0.92822	0.93216	0.93609	0.94001	0.94391	0.94779	0.95166
2.6	0.95551	0.95935	0.96317	0.96698	0.97078	0.97456	0.97833	0.98208	0.98582	0.98954
2.7	0.99325	0.99695	1.00063	1.00430	1.00796	1.01160	1.01523	1.01885	1.02245	1.02604
2.8	1.02962	1.03318	1.03674	1.04028	1.04380	1.04732	1.05082	1.05431	1.05779	1.06126
2.9	1.06471	1.06815	1.07158	1.07500	1.07841	1.08180	1.08519	1.08856	1.09192	1.09527
3.0	1.09861	1.10194	1.10526	1.10856	1.11186	1.11514	1.11841	1.12168	1.12493	1.12817
3.1	1.13140	1.13462	1.13783	1.14103	1.14422	1.14740	1.15057	1.15373	1.15688	1.16002
3.2	1.16315	1.16627	1.16938	1.17248	1.17557	1.17865	1.18173	1.18479	1.18784	1.19089
3.3	1.19392	1.19695	1.19996	1.20297	1.20597	1.20896	1.21194	1.21491	1.21787	1.22083
3.4	1.22377	1.22671	1.22964	1.23256	1.23547	1.23837	1.24127	1.24415	1.24703	1.24990
3.5	1.25276	1.25562	1.25846	1.26130	1.26413	1.26695	1.26976	1.27256	1.27536	1.27815
3.6	1.28093	1.28371	1.28647	1.28923	1.29198	1.29473	1.29746	1.30019	1.30291	1.30563
3.7	1.30833	1.31103	1.31372	1.31641	1.31908	1.32175	1.32442	1.32707	1.32972	1.33237
3.8	1.33500	1.33763	1.34025	1.34286	1.34547	1.34807	1.35067	1.35325	1.35583	1.35841
3.9	1.36098	1.36354	1.36609	1.36864	1.37118	1.37371	1.37624	1.37877	1.38128	1.38379
4.0	1.38629	1.38879	1.39128	1.39377	1.39624	1.39872	1.40118	1.40364	1.40610	1.40854
4.1	1.41099	1.41342	1.41585	1.41828	1.42070	1.42311	1.42551	1.42792	1.43031	1.43270
4.2	1.43508	1.43746	1.43983	1.44220	1.44456	1.44692	1.44927	1.45161	1.45395	1.45629
4.3	1.45861	1.46094	1.46325	1.46557	1.46787	1.47017	1.47247	1.47476	1.47705	1.47933
4.4	1.48160	1.48387	1.48614	1.48840	1.49065	1.49290	1.49515	1.49739	1.49962	1.50185
4.5	1.50408	1.50630	1.50851	1.51072	1.51293	1.51513	1.51732	1.51951	1.52170	1.52388
4.6	1.52606	1.52823	1.53039	1.53256	1.53471	1.53687	1.53901	1.54116	1.54330	1.54543
4.7	1.54756	1.54969	1.55181	1.55392	1.55604	1.55814	1.56025	1.56235	1.56444	1.56653
4.8	1.56861	1.57070	1.57277	1.57485	1.57691	1.57898	1.58104	1.58309	1.58514	1.58719
4.9	1.58923	1.59127	1.59331	1.59534	1.59736	1.59939	1.60141	1.60342	1.60543	1.60744

X	0.00	0.01	0.02	0.03	0.04	0.05	0.06	0.07	0.08	0.09
5.0	1.60944	1.61143	1.61343	1.61542	1.61741	1.61939	1.62137	1.62334	1.62531	1.62728
5.1	1.62924	1.63120	1.63315	1.63511	1.63705	1.63900	1.64094	1.64287	1.64480	1.64673
5.2	1.64866	1.65058	1.65250	1.65441	1.65632	1.65823	1.66013	1.66203	1.66393	1.66582
5.3	1.66771	1.66959	1.67147	1.67335	1.67523	1.67710	1.67896	1.68083	1.68269	1.68454
5.4	1.68640	1.68825	1.69009	1.69194	1.69378	1.69561	1.69745	1.69928	1.70110	1.70293
5.5	1.70475	1.70656	1.70838	1.71019	1.71199	1.71380	1.71560	1.71739	1.71919	1.72098
5.6	1.72277	1.72455	1.72633	1.72811	1.72988	1.73166	1.73342	1.73519	1.73695	1.73871
5.7	1.74047	1.74222	1.74397	1.74572	1.74746	1.74920	1.75094	1.75267	1.75440	1.75613
5.8	1.75786	1.75958	1.76130	1.76302	1.76473	1.76644	1.76815	1.76985	1.77156	1.77326
5.9	1.77495	1.77664	1.77834	1.78002	1.78171	1.78339	1.78507	1.78675	1.78842	1.79009
6.0	1.79176	1.79342	1.79509	1.79675	1.79840	1.80006	1.80171	1.80336	1.80500	1.80665
6.1	1.80829	1.80993	1.81156	1.81319	1.81482	1.81645	1.81808	1.81970	1.82132	1.82293
6.2	1.82455	1.82616	1.82777	1.82938	1.83098	1.83258	1.83418	1.83578	1.83737	1.83896
6.3	1.84055	1.84213	1.84372	1.84530	1.84688	1.84845	1.85003	1.85160	1.85317	1.85473
6.4	1.85630	1.85786	1.85942	1.86097	1.86253	1.86408	1.86563	1.86718	1.86872	1.87026
6.5	1.87180	1.87334	1.87487	1.87641	1.87794	1.87946	1.88099	1.88251	1.88403	1.88555
6.6	1.88707	1.88858	1.89009	1.89160	1.89311	1.89462	1.89612	1.89762	1.89912	1.90061
6.7	1.90211	1.90360	1.90509	1.90657	1.90806	1.90954	1.91102	1.91250	1.91398	1.91545
6.8	1.91692	1.91839	1.91986	1.92132	1.92279	1.92425	1.92571	1.92716	1.92862	1.93007
6.9	1.93152	1.93297	1.93441	1.93586	1.93730	1.93874	1.94018	1.94162	1.94305	1.94448
7.0	1.94591	1.94734	1.94876	1.95019	1.95161	1.95303	1.95444	1.95586	1.95727	1.95868
7.1	1.96009	1.96150	1.96291	1.96431	1.96571	1.96711	1.96851	1.96990	1.97130	1.97269
7.2	1.97408	1.97547	1.97685	1.97824	1.97962	1.98100	1.98238	1.98376	1.98513	1.98650
7.3	1.98787	1.98924	1.99061	1.99197	1.99334	1.99470	1.99606	1.99742	1.99877	2.00013
7.4	2.00148	2.00283	2.00418	2.00553	2.00687	2.00821	2.00955	2.01089	2.01223	2.01357
7.5	2.01490	2.01623	2.01757	2.01889	2.02022	2.02155	2.02287	2.02419	2.02551	2.02683
7.6	2.02815	2.02946	2.03078	2.03209	2.03340	2.03471	2.03601	2.03732	2.03862	2.03992
7.7	2.04122	2.04252	2.04381	2.04511	2.04640	2.04769	2.04898	2.05027	2.05156	2.05284
7.8	2.05412	2.05540	2.05668	2.05796	2.05924	2.06051	2.06179	2.06306	2.06433	2.06560
7.9	2.06686	2.06813	2.06939	2.07065	2.07191	2.07317	2.07443	2.07568	2.07694	2.07819
8.0	2.07944	2.08069	2.08194	2.08318	2.08443	2.08567	2.08691	2.08815	2.08939	2.09063
8.1	2.09186	2.09310	2.09433	2.09556	2.09679	2.09802	2.09924	2.10047	2.10169	2.10291
8.2	2.10413	2.10535	2.10657	2.10779	2.10900	2.11021	2.11142	2.11263	2.11384	2.11505
8.3	2.11625	2.11746	2.11866	2.11986	2.12106	2.12226	2.12346	2.12465	2.12585	2.12704
8.4	2.12823	2.12942	2.13061	2.13180	2.13298	2.13417	2.13535	2.13653	2.13771	2.13889
8.5	2.14007	2.14124	2.14242	2.14359	2.14476	2.14593	2.14710	2.14827	2.14943	2.15060
8.6	2.15176	2.15292	2.15408	2.15524	2.15640	2.15756	2.15871	2.15987	2.16102	2.16217
8.7	2.16332	2.16447	2.16562	2.16677	2.16791	2.16905	2.17020	2.17134	2.17248	2.17361
8.8	2.17475	2.17589	2.17702	2.17815	2.17929	2.18042	2.18155	2.18267	2.18380	2.18493
8.9	2.18605	2.18717	2.18830	2.18942	2.19053	2.19165	2.19277	2.19388	2.19500	2.19611
9.0	2.19722	2.19833	2.19944	2.20055	2.20166	2.20276	2.20387	2.20497	2.20607	2.20717
9.1	2.20827	2.20937	2.21047	2.21157	2.21266	2.21375	2.21485	2.21594	2.21703	2.21812
9.2	2.21920	2.22029	2.22137	2.22246	2.22354	2.22462	2.22570	2.22678	2.22786	2.22894
9.3	2.23001	2.23109	2.23216	2.23323	2.23431	2.23538	2.23644	2.23751	2.23858	2.23965
9.4	2.24071	2.24177	2.24283	2.24390	2.24496	2.24601	2.24707	2.24813	2.24918	2.25024
9.5	2.25129	2.25234	2.25339	2.25444	2.25549	2.25654	2.25759	2.25863	2.25968	2.26072
9.6	2.26176	2.26280	2.26384	2.26488	2.26592	2.26696	2.26799	2.26903	2.27006	2.27109
9.7	2.27213	2.27316	2.27419	2.27521	2.27624	2.27727	2.27829	2.27932	2.28034	2.28136
9.8	2.28238	2.28340	2.28442	2.28544	2.28646	2.28747	2.28849	2.28950	2.29051	2.29152
9.9	2.29253	2.29354	2.29455	2.29556	2.29657	2.29757	2.29858	2.29958	2.30058	2.30158

TABLE A.6. PERCENTAGE POINTS OF BARTHOLOMEW'S

TEST FOR ORDER WHEN M = 3 PROPORTIONS

ARE COMPARED

C	.10	.05	ALPHA .025	.01	.005
0.0	2.952	4.231	5.537	7.289	8.628
0.1	2.885	4.158	5.459	7.208	8.543
0.2	2.816	4.081	5.378	7.122	8.455
0.3	2.742	4.001	5.292	7.030	8.360
0.4	2.664	3.914	5.200	6.932	8.258
0.5	2.580	3.820	5.098	6.822	8.146
0.6	2.486	3.715	4.985	6.700	8.016
0.7	2.379	3.593	4.852	6.556	7.865
0.8	2.251	3.446	4.689	6.377	7.677
0.9	2.080	3.245	4.465	6.130	7.413
1.0	1.642	2.706	3.841	5.413	6.635

REPRODUCED FROM TABLE A.1 OF BARLOW, R.E.,
BARTHOLOMEW, D.J., BREMNER, J.M. AND BRUNK,
H.D. (1972). ''STATISTICAL INFERENCE UNDER
ORDER RESTRICTIONS.'' JOHN WILEY AND SONS,
NEW YORK.

THE TABLE IS SYMMETRIC IN C1 AND C2

C2	ALPHA	0.0	0.1	0.2	C1 0.3	0.4	0.5	0.6	0.7
0.0	.10	4.010							
	.05	5.435							
	.025	6.861							
	.01	8.746							
	.005	10.171							
0.1	.10	3.952	3.891						
	.05	5.372	5.305						
	.025	6.794	6.724						
	.01	8.676	8.601						
	.005	10.098	10.020						
0.2	.10	3.893	3.827	3.758					
	.05	5.307	5.235	5.160					
	.025	6.725	6.649	6.570					
	.01	8.602	8.522	8.437					
	.005	10.022	9.939	9.851					
0.3	.10	3.831	3.760	3.685	3.606				
	.05	5.239	5.162	5.080	4.993				
	.025	6.653	6.571	6.484	6.391				
	.01	8.525	8.438	8.346	8.246				
	.005	9.942	9.852	9.756	9.653				
0.4	.10	3.765	3.688	3.607	3.519	3.423			
	.05	5.166	5.083	4.994	4.898	4.791			
	.025	6.575	6.486	6.392	6.289	6.174			
	.01	8.442	8.348	8.247	8.137	8.014			
	.005	9.855	9.758	9.653	9.539	9.411			
0.5	.10	3.695	3.610	3.521	3.423	3.313	3.187		
	.05	5.088	4.997	4.898	4.791	4.670	4.528		
	.025	6.491	6.394	6.289	6.173	6.043	5.891		
	.01	8.352	8.246	8.136	8.013	7.873	7.709		
	.005	9.761	9.654	9.537	9.409	9.264	9.092		

THE TABLE IS SYMMETRIC IN C1 AND C2

					C1				
C2	ALPHA	0.0	0.1	0.2	0.3	0.4	0.5	0.6	0.7
	.10	3.617	3.523	3.422	3.310	3.183	3.031	2.837	
	.05	5.002	4.900	4.789	4.665	4.524	4.354	4.135	
0.6	.025	6.398	6.289	6.170	6.038	5.886	5.702	5.462	
	.01	8.251	8.135	8.008	7.867	7.703	7.504	7.244	
	.005	9.656	9.535	9.404	9.256	9.085	8.877	8.604	
	.10	3.530	3.422	3.305	3.172	3.017	2.822	2.550	1.987
	.05	4.904	4.787	4.657	4.510	4.337	4.118	3.805	3.137
0.7	.025	6.291	6.166	6.027	5.870	5.682	5.443	5.100	4.346
	.01	8.135	8.002	7.854	7.684	7.482	7.223	6.846	6.000
	.005	9.534	9.395	9.242	9.065	8.853	8.581	8.183	7.279
	.10	3.427	3.296	3.151	2.981	2.770	2.473	1.642	
	.05	4.787	4.644	4.483	4.294	4.056	3.715	2.706	
0.8	.025	6.163	6.011	5.838	5.634	5.375	4.999	3.841	
	.01	7.994	7.832	7.647	7.427	7.146	6.734	5.412	
	.005	9.385	9.217	9.025	8.795	8.500	8.064	6.635	
	.10	3.291	3.110	2.897	2.621	2.166			
	.05	4.631	4.432	4.195	3.883	3.353			
0.9	.025	5.990	5.778	5.523	5.182	4.591			
	.01	7.804	7.577	7.303	6.933	6.277			
	.005	9.183	8.948	8.661	8.273	7.576			
	.10	2.952							
	.05	4.231							
1.0	.025	5.537							
	.01	7.289							
	.005	8.628							

TABLE A.8. PERCENTAGE POINTS OF BARTHOLOMEW'S
TEST FOR ORDER WHEN UP TO M = 12 PROPORTIONS
BASED ON EQUAL SAMPLE SIZES ARE COMPARED

M	ALPHA				
	.10	.05	.025	.01	.005
3	2.580	3.820	5.098	6.822	8.146
4	3.187	4.528	5.891	7.709	9.092
5	3.636	5.049	6.471	8.356	9.784
6	3.994	5.460	6.928	8.865	10.327
7	4.289	5.800	7.304	9.284	10.774
8	4.542	6.088	7.624	9.639	11.153
9	4.761	6.339	7.901	9.946	11.480
10	4.956	6.560	8.145	10.216	11.767
11	5.130	6.758	8.363	10.458	12.025
12	5.288	6.937	8.561	10.676	12.257

REPRODUCED FROM TABLE A.3 OF BARLOW, R. E., BARTHOLOMEW,
D. J., BREMNER, J. M. AND BRUNK, H. D. (1972).
''STATISTICAL INFERENCE UNDER ORDER RESTRICTIONS.''
JOHN WILEY AND SONS, NEW YORK.

Author Index

Subject Index